SCIENCE AND
CONTEMPORARY SOCIETY

LUDWIG F. AUDRIETH

MICHAEL CROWE

FARRINGTON DANIELS

HERBERT FEIGL

ERWIN HIEBERT

RICHARD McKEON

PHILIP MORRISON

ELIZABETH SEWELL

JOHN E. SMITH

SCIENCE and Contemporary SOCIETY

FREDERICK J. CROSSON

Editor

UNIVERSITY OF NOTRE DAME PRESS
NOTRE DAME — LONDON

Copyright © 1967 by
University of Notre Dame Press
Library of Congress Catalog Card Number 67-22148
Manufactured in the United States of America

TABLE OF CONTENTS

Editor's Preface — vii

Foreword: Looking Backward — xi
Milton Burton

A CENTURY IN RETROSPECT

1. Science and Literature
   Elizabeth Sewell — 3

2. Scientific and Philosophic Revolutions
   Richard McKeon — 23

3. Thermodynamics and Religion: A Historical Appraisal / *Erwin Hiebert* — 57

4. Science a Century Ago
   Michael Crowe — 105

PROSPECTS AND CHALLENGES

5. Contemporary Science and Philosophy
   Herbert Feigl — 129

6. Science and Religion: A Reappraisal
   John E. Smith — 155

7. Science and International Affairs
   Ludwig F. Audrieth — 175

8. Science and Human Welfare
   Farrington Daniels 193

9. Science, Education, and the Future of Mankind
   Philip Morrison 213

   Contributors 243

   Index 245

PREFACE

With the end of the Civil War in April of 1865, the now-again United States began to assess its resources for the years of building and growth ahead. In the same year, indeed within a few months, the College of Notre Dame du Lac made a tiny but symbolic contribution to that future in awarding its first degree in science. For science and its offspring, modern technology, were to play an increasingly important role in the cultural and socioeconomic development of our society in the next century.

In 1965 the University of Notre Dame, now become an important center of scientific training and research, celebrated a Centennial of Science in commemoration of that long-ago event. It seemed an appropriate time to take stock of how the relations between science and society had been transformed, indeed how science and society themselves were transformed by those changing relations. The retrospective assessment suggested a prospective one: what were the outlines of the future, of the next hundred years? This book is the record of that assessment and vision as it was presented to a large and distinguished audience by a group of equally distinguished scientists, historians, poets, and philosophers.

Perhaps the most remarkable theme to emerge was the general agreement that the thesis of the "two cultures" was an anachronism, that the supposed opposition between the sciences and the humanities was in fact based on an outmoded conception of science. The first and last essays in this volume, written respectively by a poetess and a physicist, discuss and indeed exemplify this theme clearly and brilliantly. Part of the reason for the change in conception is examined by an eminent philosopher of science who was once a prominent protagonist of the logical positivist notion of science, perhaps the paradigm case of the older and now anachronistic view.

Conversely, that understanding of the humanities and of religion which defined them as opposed to the sciences, which conceived of the sciences as antihumanistic and antireligious, is vigorously repudiated by the humanists and religionists. Perhaps a sign of this change is the two essays by historians of science, practitioners of an "in-between" discipline which did not exist a century ago.

It became clear, then, that what is needed is a radical rethinking and renovation of our educational curriculum and the manner in which the sciences and the humanities are introduced along the stages of formal education. As one of the contributors put it, we must go back and "water the subsoil," go back and think again about the basic attitudes and presuppositions with which we form the minds and personalities of the young people of our society. Our Symposium concluded not with a resolution of these issues but with a clarification of the problems we face. It ended with a challenge for the future and a hope that having seen more clearly the shape of these problems we will be better able to move toward their resolution.

Acknowledgment for multifarious assistance and encouragement must be made to a great university president, Reverend Theodore M. Hesburgh, C.S.C.; to a distinguished humanist, educator, and human being, Doctor George N. Shuster; to the Deans of the Colleges of Science and of Arts and Letters at Notre Dame, Doctor Frederick Rossini and Reverend Charles Sheedy, C.S.C.; to Professor Milton Burton, Director of the Radiation Laboratory and to his Centennial Committee; to Dean Thomas Bergin of the Center for Continuing Education at Notre Dame; to numerous others whose generous cooperation made everything easier; and to the National Science Foundation for financial assistance with the original Symposium.

Special thanks are due to Professor Harvey Bender of the Department of Biology, co-chairman of the Symposium, to Harriet Kroll of the University Press, to Marjorie Vanderbeek and Norma Davitt for expert secretarial assistance, and last but not least to my wife, whose loving care of me made my care of the book possible.

Frederick J. Crosson

FOREWORD: LOOKING BACKWARD

Milton Burton

> *The advantage of being old is that you can predict the past.*
> —George N. Shuster

THE DEVICES OF HISTORY

In conversation with the distinguished educator George Shuster, I once remarked that the most convincing historical reviews, at least of fairly recent events, are by the elderly, who convey a sense of immediate participation, and that the most useful prophecies are by the young, untrammeled by the prejudice of experience. I suggested also a converse proposition: that the young man might present history with impersonal objectivity and that the older man might see the future by looking backward—not necessarily from this moment in time but from some future moment that he might arbitrarily choose. For imagination he could substitute experience, and for awe of the future he could substitute acceptance of the past. Dr. Shuster's very quotable remark and this preview of science are respectively the immediate and remote consequences of our exchange. In seven integrated books Proust recaptured the past; the intent of this foreword is to *recapture* a very limited fraction of the future.

SCIENCE AND TECHNOLOGY

When I look back to my high-school days, I find that, although my major interest was always in science, the teachers whom I best recall and whose names I remember were those who gave me a sense of quality, those who taught me Latin and English and history. I remember the teachers of science less vividly, and I have no recollection at all of any of the teachers of mathematics. Perhaps, deterioration had already begun to set in a half-century ago. In relatively recent times teachers were hired to coach one of the sports and assigned to the teaching of some branch of mathematics because they could there do a minimum of harm—or so it was thought.

But technology has become vividly alive in the last quarter-century, and attitudes toward science and math have changed. The men of science are needed and encouraged because the world wants the fruit of their efforts. The high-school teachers of math no longer go to "coaching clinics" in the summer; they attend "refresher courses" in mathematics.

When I attended college and the university, it never occurred to me that one branch of learning could be counted higher than another. In fact, it was the humanists who were the best known among the scholars and who set the tone of the university. But because all teachers of those days were notoriously underpaid, there were no discomforting disparities, no false standards. Suddenly, however, technology has contributed wealth, made its demands, and offered its rewards. The scientist finds himself set apart, more completely a specialist, and, if he knows history, wonders at what happened.

Toward the end of this volume Farrington Daniels emphasizes technology as a product of science, rather than as science itself. According to currently accepted notions, the reading public can appreciate science only in terms of technology, in terms of accomplishment that produces some

Foreword: Looking Backward

type of hedonistic satisfaction. In our weekly news reviews, the title of the section that deals with science varies with the times; it is Science less frequently than it is Science and the Atom or Science and Medicine or Science and Space. By contrast we have Art or Music or Theater or Literature. A section on Music and Medicine might, on occasion, be timely; the reader would quickly recognize that a special point is being made. By contrast, a section on Science and Technology would evoke no reaction at all. It is accepted that the two *always* belong together.

The theme can be examined endlessly. Congress understands that science is important and indeed wonderful but must be reminded repeatedly, in terms of technological fruits or "spin-off," that it is really worth support. Accomplishment in the life sciences is financially encouraged, not because it may be intrinsically important for man to know himself, but because of urgent concern about cancer or muscular distrophy or heart disease or some other affliction which may have struck a public figure. This is the technological age.

Part of our present misunderstanding of the role of technology in our society is the fact that science has become more and more dependent on technology. Should an astrophysicist want to learn more about remote space, he must extend his efforts even beyond the massive optical machinery of the modern observatory. A radar telescope 200 feet across, mounted on tracks and a framework of girders, so controlled that it automatically remains locked in on some remote object in space, is a scientific necessity. Astronomical observations by optical spectroscopy demand space platforms so that the data recorded need not be affected by the artifacts introduced by our own atmosphere. The offspring of high-grade, and very expensive, technology now becomes science. Curiously, this expensive reality does not diminish the probability of adequate support. Rather, it increases such probability. Large industry becomes involved in the production of science. In our affluent society, financial profit and the employment of many men become goals in themselves.

We are concerned with a circular phenomenon. Need has created technology. Science has improved technology. Technology (as in problems of pollution) has presented science with valid problems. Science has grown so greatly that it itself makes demands on the most skillful of the technologists who, in turn, employ the most sophisticated achievements of the theoretical scientists for the solution of their own immediate problems.

Is it then our conclusion that, in the next 100 years, science and technology will become so intertwined that an effort at distinction is purposeless? Are we to accept the suggestion (by an occasionally provocative educator of our times—I have one specifically in mind) that a scientist is really and merely a gadgeteer?

Questions such as these are not trivial. If we are to project the future, we must understand our past. The ingenuity of the technologist has created an affluent society and the related problem of an unprepared leisure class. In the past we have accepted the idea of "shirt-sleeves to riches to shirt-sleeves in three generations" as a consequence of the weakening of the fiber (moral or intellectual) of the affluent generation. An entrenched aristocracy and the rules of entail have delayed the process in other countries. In our own, the process has become diluted (and perhaps ended) by the spread of affluence and by a growing understanding (on the part of the increasing, increasingly wealthy class) that special opportunity creates special responsibility. A puritan mystique now prevails. The call of duty is heard and an aristocracy of wealth is becoming an aristocracy of leadership.

A new technology ultimately demands less and less of more people. Man has and will have increased opportunity for contemplation and creation of beauty in numbers, in science, in the world around us, in art, and in language. Because moral fervor will no longer be able to expend itself on service to a deprived minority and because education will penetrate deeply into our classless society, the leisured will seek outside their intellectual or aesthetic preoccupations for meaningful (not wholly recreational) physical endeavor in response to more and more automated technology.

Life may be described as a process which adjusts itself to its environment so that, as the environment changes, the process changes in such a way that it remains equally efficient or becomes more so. (In death, the process becomes less and less efficient and ultimately terminates.) A living process involves a mechanism which compensates automatically for its own deterioration by creation of another mechanism that can perform the same process. With so broad a definition, life is not restricted to organic matter. It is even possible that the technological, intellectual gods of the new leisure class can create "advanced vehicle" factories (as an example) which are programmed to interact with other types of factories so that in their maturity they may produce new factories of the class for which they are programmed. The program itself would depend on the intellect, the imagination, the moral sense, and the fervor of a class which is reluctantly leisured and which, hopefully, may survive the class of machines it has created so that it may insert improvements or make some more desirable replacement. I write these words merely to indicate that a prediction that we will be able "to create life" is fatuous indeed. The creation of life is a realizable process right now —if we define life in appropriate terms and are willing to expend adequate funds and effort.

It is strange that in a foreword to a book on science, technology should be so stressed; and it is when we talk about creation of "life." However, it is mainly through an awareness of technology that science (as well as its uses) can be understood. Ask a man to predict the advances of science and he will look at present science and predict the technology of the future; the more courage and imagination he possesses the more into the future will he make his predictions. By contrast, prediction of scientific advances, except on essentially trivial matters, is not possible. One can say, as an example, that we will learn how to solve certain mathematical equations (e.g., problems of quantum mechanics) which may eliminate the need for certain tedious experiments. It seems almost unquestionable that the life scientists will break genetic codes and make planned modifications at will, synthesize complex proteins, and learn the elaborate details of perception, response, memory, and thought. And we can expect that investigations of particle and sub-particle physics will lead us to sub-sub particles. Dean Swift captured the same idea when he meditated on the louse—without resort to scientific pomposity. We can make minor projections endlessly if we make them only into the immediate future.

The real lesson we can learn from the past about science is that we can neither imagine nor predict it very far in advance. However, we can think profitably about the *sources* of science in the future. In 2065 the problem of science and technology will no longer exist. Technology will be sufficiently advanced to meet all needs. The new problem will become that of man and the uses of technology.

In recent years, many scientists have been preoccupied by essentially ecological problems. "What happened to the dinosaurs?" The disappearance of the dinosaur is not a unique phenomenon. There have existed entire classes of animals the history of which appears to have terminated abruptly. "Why is it that in some remote past the tropical and frigid zones were differently located?" Answers relating such effects to the migration of our magnetic poles, to the strength of our magnetic field, to the action of external proton bombardment of our planet, and so on, are beginning to emerge. They are so enormous in their implications that

Foreword: Looking Backward xvii

it is doubtful whether the scientist may ever do more than ask the technologist to assist in their verification. It strains the imagination to believe such a possibility but, because we do think about it, we must accept the probability, however small, that the scientist and the technologist working together will ultimately concern themselves with the problem of survival of our world in our very curious universe. However, it is utterly unlikely that this problem will be attacked in the next 100 years. On the other hand, the more immediate ecological problem of maintenance of life on a planet on which humans are exhausting its natural resources may become of predominant importance in the next 100 years and terminate both the problems of an affluent society (of scientists, artists, humanists, men of letters, technologists) and the affluent society itself.

It is an occasional assumption that the relationship between science and technology is self-evident and that in this respect (see the scientist as a gadgeteer) science is different from the arts and the humanities. However, the truth is that all culture both makes demands on, and enriches, technology and is enriched and made more vital by it. Consider art and the architect, the performing arts and the theater, or, in one close group, painting, sculpture, the museum, and architecture. Or note speed-reading and electronics or the learning of language and the language laboratory. The existence of recording and reproducing devices has not only improved the teaching and learning of language but has also greatly affected the nature of the subject learned as well as the quality and character of international communication.

LEARNING AND ITS USES, 2065

And now I am young again, looking backward from a future which has become my present. Can it be true, I inquire, that in a distant past people confused themselves by classifying learning as to quality (in addition to type)?

Did they really think a man could be learned without thought of moral duty? Were they so in fear of individual survival, were they so conscious of their own persons and of what they construed to be their personal needs or advantages, that they budgeted the uses of intellect and argued the relative merits of achievement in the various fields of intellectual endeavor, that they looked upon technology as a sort of intellectual prostitution and deprecated any branch of knowledge which served technology?

What was the idea? Was man so limited in intellect, in powers of expression, in capacity for feeling that he was restricted to attainment in one of these areas and perforce had ignored or neglected the others? To be sure, we are not yet complete, but at least we regret our ignorance and are happy to contribute to those varieties of technology which make 2065 such an intriguing and exciting year in which to live.

SCIENTIFIC LITERATURE, 1965

One of the less questionable advantages of moderate age is that you can look backwards at your innocence, your early anticipations and limited expectations, and your surprises; and, in terms of such experience, if you are still alert and young enough to be imaginative, state some views about the future which are startling and yet make a considerable degree of sense.

In the last forty-five years I have been in the midst of science and, for some five to ten years before that, aware of it. I remember that at the age of ten or thereabouts I was profoundly affected by a book by Soddy (or was it Ramsay?) on radiochemistry (or radiophysics—not *nuclear* chemistry for the idea had hardly been born) and knew from then on that my career was to be in science. So, I am well grounded in my prejudices and in my obstinacy, for they go back to an early time, to a time when one could write a book which might charm a ten-year old and, at the same time, tell all worth recording about a complete area of science in a language adequate to the requirements and the curiosity of a graduate student.

In about the same period and probably within the five years immediately thereafter, I had an experience in a doctor's waiting room which remains with me so vividly that I am convinced that it was real. Just as now, they had news magazines in waiting rooms in those days—and, because tradition is powerful and based on real origins, they were probably old magazines. I remember reading in one such magazine (could it have been the *Literary Digest*?) of Harkins' idea of the compound nucleus, a very profound idea indeed for which he received in his lifetime very little credit. Memory is a strange thing. Have I confused two waiting rooms? Was it ten years later than I remember? Were both I and the world readier for profound ideas? How then was such an amazing idea lost and momentarily (in the time of the universe) forgotten before it became an integral part of our most fundamental ideas about matter? The answers to such questions are unimportant. What is important is that a child in those days could be exposed to ideas profound enough to titillate the most advanced intellects. Science, the knowledge of science, were still, in some degree of completeness, achievable goals.

The history of science is charming indeed. Davy was able to convey to Faraday the total knowledge of electrochemistry then available in the world. No one can seriously doubt that anyone in Faraday's old age knew anything about electrochemistry and related subjects which Faraday himself could not quickly assimilate by reading, by reading which did not interfere with the progress of experiments in his laboratory. I have read, or perhaps been told the story, that when Wilder Bancroft was a very young man, he prepared himself for a career as a chemist by reading all the back issues of the *Journal of the American Chemical Society* (and perhaps its predecessor whatever its name happened to be). Of course, when I joined the American Chemical Society some forty-five years ago, such a task would have been much more formidable—so that it never occurred to me to attempt it. But I did read every new issue of the *Journal* from cover to cover (for what good purpose I cannot recall), and I did read *all* the items in the then less formidable *Chemical Abstracts*. As I grew older, more sophisticated, and busier my appetite became more selective, and I found myself reading a proportionately lesser amount of a constantly expanding literature. Today, I depend more and more on the spoken word from my students and my associates and the colleagues I encounter at the scientific meetings I attend. There was a time when attendance at all such meetings of any degree of importance was a physical (if not a financial) possibility. Now, such times are past. My curiosity remains full blown, and the number of meetings which titillate my fancy grows steadily but, of course, I attend only a very small fraction of them, and not nearly so effectively as I attended the occasional meeting of my early days. Then, I was thoroughly aware of the background of the work reported and discussed. Now, I know that there is a background, the minutiae of which I no longer attempt to grasp.

What is the lesson from this brief recital of one phase of *my* scientific experience? Only, that the literature of science is too overwhelming for utility. One skips lightly over facts and misses profound ideas—or grasps them, if at all, only partially and thus not really—for if one really studies one item, he may miss the existence of ten others, or more. Of course, the specialist, the very, very true specialist, can limit himself to one brilliant facet of the jewel of science and, looking into it singly and closely, may see everything else by reflection. But the person and the operation are rare indeed and the flood of printed words may submerge and overwhelm even him.

Foreword: Looking Backward

What I have learned then in approximately fifty years about the record and recording of scientific progress is only that it has become a vast, vast enterprise and that its utility is becoming marginal.

In a spoken foreword to the Symposium which this book surveys, I made the point that scientists do understand that there are kinds of knowledge which cannot be measured but only felt. Also, scientists do have a degree of skepticism regarding the things they measure and presumably test and verify. Their skepticism extends both to the inferences developed from such measurements and to the measurements made to test the acceptability of their profoundest ideas. The difficulty with scientific knowledge goes all the way back to the meaning of knowledge. However, we may stop short of the elementary profoundity and address ourselves to the simpler question of the worth of our inferences based on the most reliable of measurements.

Symbolic logic was not a subject which attracted the attention of any great body of scientists fifty years ago. However, in the present day, students in attendance at summer mathematics institutes may be exposed to symbolic logic while still in their early teens. It is offered as a terminal mathematics course in some institutions to students who do not intend to go further in their mathematical explorations in their chosen careers. Presumably, the philosophy of such an offering is that *perhaps* the future lawyer may thus learn to test the validity of his argument or his plea or that a future judge may thereby be given the tools to test the difference between brilliance and proof. Or that the future physician may learn the difference between good fortune and wisdom—and learn what wisdom is. The idea is easily extended. The mathematician wishes, if he can do no better, to cast good seed.

Nevertheless, symbolic logic does have direct utility for a scientist. If he employs it rigorously, he can state both his conclusions and their degree of acceptability. He can delineate his knowledge and thus express his ignorance, and do so in the form of a mathematical symbolism which has rigorous meaning. Symbols, properly defined and understood, have singularity of meaning. They occupy little space and do, in principle, make possible the transmission of meaning.

The real problem of communication in science is the transmission of meaning, just as in science itself the real problem is the development of meaning. The student of science, the explorer of the literature of science, loses himself all too frequently, not in a barren waste-land, but in a lush jungle of information relieved occasionally by the existence of a cultivated patch in which he sees order, symmetry, purpose, and occasionally product. Unfortunately for the explorer, these refreshing gardens of clarity do not contain all the wealth of the jungle. Much remains to be unraveled, to be examined, and to be interpreted. Because the number of searchers does not increase as rapidly as does the lush growth of the jungle and because even the best of explorers is finite and grows weary and discouraged, intrinsic worth is lost almost as rapidly as it is produced.

How many of us have read an article, stumbled over the conclusion of its author (or, in the style of the present day, multiplicity of authors) and said to ourselves "What the devil does he mean?" Even worse is the experience with an author who fails to indicate purpose and related content, and stops short of interpretation of something really new. And then the question becomes "What the devil does *it* mean?" Well, among other things, "it" can mean bad techniques or bad observations and, failing the effort of a determined author convinced of his facts and hidden objectives (objectives, unfortunately, hidden sometimes from himself), "it" becomes lost in the limbo of forgotten facts. When the facts are rediscovered by a person who knows *why* he wants them, they first become real and useful, and the original discoverer bemoans the existence of a world which

Foreword: Looking Backward

seems to have ignored him, deliberately and contumaciously. The one occasional result is the addition to the already muddled literature of science of the statement of a prior claim, the seed of polemical literature.

In any forecast of the future, even of science, it is a mistake to ignore the impact of human vanity or of human limitations or of human needs. But it is a mistake also to misinterpret or misevaluate any of them. Need is affected by possibilities so that it is the possibilities of the future which we must first determine. Those possibilities are intimately entwined with immediate and future probabilities. Consequently, it is the latter which we now examine.

Scientists are no longer adequate repositories of scientific knowledge. That day, if it ever existed, passed while I was still a very young man. The repositories of knowledge are the libraries, the notebooks, and microphotographs of notebooks, supplemented by file cards, occasional tapes and drums, and vague memories. Repositories, we know, are not enough. Retrieval of information, assessment of meaning and of value, and transmission of meaning are all-important. Retrieval of meaning in a useful time is the important objective. And what is a useful time?

"Useful time" depends on the value of the person who uses it. Ideally, the research person, evaluating and interpreting his studies (and his facts), in their turn developed on the presumption of an adequate knowledge of the pertinent literature, should have access to pertinent facts and ideas *at once* to a degree determined only by his own limited ability to employ the information he receives. The words "at once" also require definition. For most people, "at once" is related among other factors to reading speed, to mental flexibility, and to quickness of perception and assessment. A time like twenty seconds or twenty minutes would delight most of us who are bogged down by the need for literature search.

SCIENTIFIC LITERATURE, 2065

If, now, I imagine myself to be a young scientist of the year 2065 reviewing the events of the previous 100 years, I would make the statement: It is hard to believe, except for those historians of science who browse in old and forgotten libraries, that there was once a vast literature of science published in what appears to be serial form in thousands upon thousands of scientific periodicals appearing monthly, semi-monthly, and even weekly. This so-called open literature was supplemented by an approximately equal number of reports destined for locked files, for Congressional Committees, for the patent attorney, or for oblivion. Furthermore, there were journals devoted to publication of "letters" —considered to be so important that society could not endure the usual five- or six-month delay involved in regular publication. There were many journals of abstracts of the current literature, so extensive that no one could read all the current abstracts of his own field. There were journals of titles, so that one might *perhaps* at least become acquainted with titles if not with subject matter. There were journals of "key words" which might permit someone, were he so inclined and had the time, at very least to become acquainted with the numerous new titles of works in his specific area of interest.

Contrast this uncivilized conglomeration of scientific "knowledge" (perhaps we should say "ignorance") with the order of our present day [2065]. A scientist may think and produce without being overwhelmed by, or contributing to, the overpopulated, so-called scientific literature. In the first place, he now no longer need express himself in the unsuitable language of poetry or of the spirit such as English, or German, or French, or Latin, or, even more unsuitably, in the beautiful languages of felt ideas, such as (I am told) some of the ancient Chinese. The precise symbolism of mathematics is alone meaningful for us. We have ready access to all measurable and logically interpretable facts.

Foreword: Looking Backward

The scientists of 1965 did not imagine a world in which there was no scientific literature. How could they conceive of a time like ours when the push of a button would produce on our desks, within our vision, both the existent conclusions on any body of subject matter and, to the degree and detail we want, the facts and the logic upon which these conclusions are based. They could not imagine a time like ours when we program our computers to answer such questions as "If a, b, c, d is true, does it follow that X or Y or Z is true?" They could not even understand the possibility which now haunts us of getting a useful answer to the question "If a, b, c, d is true, what then is true or what should we next do to determine something true and useful?" Of course, some of us are haunted by the question of what would be our own usefulness in a world presumably dominated by the computer. Let me assure you readers of 2065 that scientists need not fear. The computer is able to do only what we put into it. What man has is imagination. Thus, we must neither expect nor fear that the next 100 years [2065–2165] will see the elimination of need for scientists or the diminution of our role.

In the period from 1865 to 1965 it became evident that the ancient notion of men who could be encyclopedists had been reluctantly abandoned. In 2065 we can foresee a future in which the encyclopedia of knowledge and the means of its evaluation are both entered into computers. We can see a future limited not by devices, but by imagination, by the ability for intellectual employment and by the moral strength of humans to employ their leisure and to cope with the power with which technology daily endows them.

A Century
In Retrospect

1

SCIENCE AND LITERATURE

Is inexhaustible that which is real.

Elizabeth Sewell

One hundred years is a splendidly arbitrary period to celebrate. It marks perhaps nothing more than mankind's love affair with the number 10, since some genius invented the digit "zero" and we took to Arabic numerals. Nevertheless, I feel a real gratitude to the University of Notre Dame that it was in 1865 that they began conferring degrees in science and so occasioned this book. Had it been fifteen years before, in 1850, I with my allotted subject of science and literature would have been irretrievably stuck with Tennyson's *In Memoriam*. As it is, I can begin instead, when we get to the center of our subject, with two real winners and without undue cheating over exact dates (though I shall do a little cheating of this kind when I need to). These are Victor Hugo's essay on Shakespeare in 1864–1865 and Walt Whitman's "Passage to India," which in 1869 was submitted to a literary magazine—and rejected. I confess I was also much tempted by another work of this exact date, 1865, *Alice's Adventures in Wonderland,* as a possible starting point for my remarks here; however, I resisted that temptation.

What I want to say has, to help it along, a motto and a shape. The motto stands above: "Is inexhaustible that which is real." It has a special and slightly comic suitability because the thought—and a wonderful thought it is—is that of a scientist, Michael Polanyi; only then I put it into that particular shape, so it is the mixed production of a scientist and a poet. The shape of what I have to say is best thought of as two big curves. Think of them as two rainbows which intersect near the foot. One is big, and that is our hundred-year span of science and literature. The other, which is much smaller, has a span of twenty years and represents my own experience in this matter. In between the two, as we pass through in mid-course the small space where the curves intersect, I hope to say a few vituperative words about education even though that is not my official subject at all.

The motto gives us the principle of what we keep and look at and what we throw away in our subject matter. The shape gives us a moving progression other than a dead straight line. Both, I hope, will save us from that most frustrating thing to instructor and student alike, a survey course.

We had better glance, first of all, at what we are landed with, or at least I am landed with, as regards range and field. "Science and Literature, 1865–1965" is, as is plain, so crazily enormous that I decided to do no research on it at all but simply to pull up what I knew and to hope that that would be all right. Take science since 1865, and think about that. Up comes pretty well everything, from medicine to military science, cybernetics, big-bang versus steady-state, quasars, lasers, all modern psychology and psychiatry, Schliemann and Mycenae and all that (and only recently the English *Times* reporting a wonderful find of gold cups on the site of Nestor's palace at Pylos, and the scientist in charge saying, "Well, well, so Homer was right after all"), and not only Schliemann and Mycenae but also Spemann and organizers, protein crystallography, systems engineering in rockets, the sex life of bacteria, atomic particles which are right-handed and left-handed, anti-matter, and so on. Turning to literature in the same century, there is the novel, drama, poetry, criticism, through to the movie and the comic strip, both of which belong here with their more august counterparts; and if we turn to personalities, we range from the three already mentioned, Tennyson, Hugo, Whitman, to any three names we may care to mention among our contemporaries who could in our opinion take on those notable pugilists. And for countries, since the assignment was in comparative literature, there are France and Germany and the United States and England; there could be, of course, plenty more, but these are all that I know anything about, even approximately.

So, there are two threads, that of science and that of letters. They go forward together through our hundred-year span, and we have to look at the nature of the relationship between them. For they are separate, and yet, and yet . . . we find ourselves wanting to qualify that statement as soon as made. We shall start by seeing, with the help of our motto, what we can throw out—those relations between science and literature which are exhaustible and, therefore, not real. There seem to me to be four of these.

The first is, if unreal, very endearing. It consists of the scientists who dash into poetry and the poets who dash into science; and there are more of each than we might suppose. In the first group, within our time span, I think of Alfred Russell Wallace and Sir Ronald Ross, and again there is that haunting mathematician and symbolic logic man murmuring about Jabberwocks and Father William. Of the poets who dash into science, the two most notable are really out of my reach by reason of their dates. These are Goethe, of course, and Poe, who expounds the big-bang theory in *Eureka* in 1848 (where did he get it from, one wonders?). Well within our field, however, are those late nineteenth-century poets in France who developed a passion to transform language into the pure emptinesses of mathematics or of music: Mallarmé, whose work was called, by Valéry, a starry algebra, and Valéry himself, who at one point in his life gave up poetry altogether to devote himself to mathematics for some years. It is the more interesting to find Mallarmé saying of his own work, "Mon oeuvre est une impasse." Exhaustible, in other words. We must move on.

The second impasse we meet in the relations between science and poetry is where hostility between the two pursuits is exhibited, or a disjunction on principle, as for example in Professor I. A. Richards' essay *Science and Poetry* of 1926. There are a number of poems on the hostility theme, all of them unsatisfying. Poe's sonnet is too early again, but we might recall *Locksley Hall Sixty Years After;* and my own memory has retrieved something by Walter de la Mare—startling in so true and lovely a poet—where Poetry appeared as a beautiful lady and Science as a . . . *could* it be a pig? I went and looked it up, and there was "Shaggy Science nosing in the grass," Science that "snuffled . . . squealed . . . grunted." My memory had served me all too well. There is no way through for poetry (or science) here.

The third relation between science and literature which also, I believe, turns out to be exhaustible is when literature chooses to treat science as a problem and a conflict of some sort is set up or assumed as the center of the action. This mainly applies to drama, whose essence is conflict, as we believe, and so it ought to work; it applies to some extent to novels as well. Yet here too there is a rarity of good work. As specimens of this type of relation I summon up Bourget's novel *Le Disciple* of 1889, Sir Charles Snow's early novel *The Search,* and a couple of plays I have read or seen, *The Flashing Stream* by Charles Morgan and one, more recent, called *The Burning Glass* whose author escapes me. (I can think of only one novel and one play, both excellent in a rather off-beat way, which deal with the real relation between science and the arts; but neither treats science as a problem, and I shall hold them for the moment.)

The fourth and last of the exhaustible relations between literature and science is where science becomes the object of fantasy—not imagination, which we shall come to shortly, but sheer fantasy. This heading might include all science fiction, from Jules Verne and Wells to Fred Hoyle; the moral allegories such as C. S. Lewis' *That Hideous Strength;* and technological fantasies of many kinds, from Čapek's *R.U.R.* to the Daleks who are now so prevalent in England and who are rather nice. This is also the world of popular horror movies and comics. The fantasies swing, it seems, somewhere between horror and humor, those curious twins. Even in the would-be nightmares of technology, *Brave New World* or the vastly inferior *1984*, there is an odd sense of collusion, as if we were enjoying making ourselves shudder at our own inventions, an enjoyment similar to that we get from the parallel jokes that tackle the same subject matter. Should we perhaps remember here Charlie Chaplin in *Modern Times* and an echo of that same nightmare-joke in Ralph Ellison's *Invisible Man*?

It seems to me that in our hundred years there were two big waves of jokes about science which we might observe. The first consists of the Darwinian jokes which had begun rather crudely with jokes about somebody's grandfather being an ape, and went on, and on.

> A lady fair of lineage high
> Was loved by an ape in the days gone by. . . .

So the song says, continuing its tale:

> He bought new ties and he bought dress suits,
> He crammed his feet into bright tight boots,
> And starting life on a brand-new plan,
> He christened himself "Darwinian Man."
>   But it would not do. . . .

Follows the sour conclusion that a man, no matter how well behaved, at best is only a monkey shaved. This is

W. S. Gilbert, and *Princess Ida,* and 1884, quite a long time after the publication of *The Origin of Species.* The joke is still around, for I remember my sister giving my father, who was a biologist, a little china ornament representing a monkey seated, evidently deep in thought, on a big volume labeled *Darwin.* The other wave of jokes I have in mind deals with electronic brains. A pair of examples: two white-coated technicians poring over the tape which a large computer behind them has just emitted, the one announcing to the other, "It says *Cogito ergo sum";* and a birthday card I bought recently for a scientist friend, depicting on the outside a vast computer and the legend, "I am glad to be able to send you a birthday card printed by computer and the marvels of modern science, completely foolproof, nothing can go wrong," while inside the card is printed "Merry Christmas."

We are uneasy, aren't we? And the interesting thing is that if we follow these two joke-horror fantasy themes, we emerge into real literature and real imagination, exchanging the exhaustible and unreal relations between science and literature for something quite different. The Darwinian theme leads into the great Rougon-Macquart cycle of novels by Zola, where the scientific theory—heredity and environment with technology thrown in—becomes the dynamic myth directing such works as *L'Assommoir* or *Germinal* or *La Bête humaine.* The other theme leads into John Wain's admirable piece of 1956, "Poem Feigned to Have Been Written by an Electronic Brain," in which the computer, gone wrong or gone mad, tells its appalled human minders that truth is not with it, the machine, but with them, twined somehow into the pulses and earth of the human body. (Henry Reed's "Naming of Parts," one of the very few good poems to come out of England in the last war, says somewhat the same thing.)

We have here, first in myth and second in the simple affirmation of the human frame of body and mind, two great methods of the imagination—perhaps the only two for dealing with matters too enormous for our capacity. Such would include the enormities with which we have been confronted by science and technology within our lifetime. There are, it seems to me, two terminal points here which we are in duty bound to look at. The first is Auschwitz. One can understand why Rolf Hochhuth wanted to write his play *The Deputy* in verse and why, in dealing with the terrible figure of the camp doctor, he reaches out towards the resources of myth, resources which go back not only through Zola but Kafka's *Penal Colony* and Freud and Marx too, great myth-men, and which suggest that *Paradise Lost* (remember that "devilish Enginrie"?) might be among the works of reference for this first terminal point of our technological age. The second terminal point is Hiroshima. And in the literature about this we may see that other method of the imagination just spoken about, the affirmation of simple humanity, the human being as sentient and conscious organism, for instance, in John Hersey's *Hiroshima* and again in that remarkable movie *Hiroshima mon Amour,* which says that the suffering of tens of thousands is not in itself more than the suffering of one single human being, the anguish of one French girl in Nevers as absolute as the holocaust in the Japanese city—that is, if I read it aright.

We had to look at both of these. But now we may exorcise them and move on to something essentially human and blessed, the function of the imagining intellect which both conceives and constitutes the real, inexhaustible relation between science and literature which I shall now include in the term "poetry," since poetry will mainly occupy us from now on. It is good, therefore, to use by way of exorcism a poem of a hundred years ago so that we can shift back to our beginning (actually it is slightly more, 1857—I am cheating a little)—a poem in which Longfellow suggests what it may mean to be a scientist:

> So he wandered away and away
> With Nature, the dear old nurse,
> Who sang to him night and day
> The songs of the universe;
>
> And whenever the way seemed long
> And his heart began to fail,
> She would sing a more wonderful song,
> Or tell a more marvellous tale.
>
> So she keeps him still a child,
> And will not let him go,
> Though at times his heart beats wild
> For the beautiful Pays de Vaud . . .
>
> And the mother at home says, Hark!
> For his voice I listen and yearn;
> It is growing late and dark,
> And my boy does not return.

What a beautiful disturbing image for poetry to choose for a scientist (the poem is for Louis Agassiz on his fiftieth birthday). Here our two threads of science and poetry meet and join, in what happens in the mind.

Take two little sample knots, or nerve ganglia, as instances of how inseparable in reality the two threads are. The first centers around Sainte-Beuve, that man of letters who claimed Bacon as his model for literary criticism, adding that this craft of his is akin to the taxonomic skill of the botanist Jussieu or the comparative anatomy of the great Cuvier. In a particularly interesting passage in the *Premiers Lundis* of 1886 Sainte-Beuve picks up the, for him, significant chance that Cuvier on his election to the Académie had, as tradition demands, to pronounce the eulogy of the dead man whose seat he inherited, in this case the poet Lamartine. Of this Sainte-Beuve says, "Himself a man of genius who has arrived at those heights of science where it is scarcely distinguishable from poetry, M. Cuvier was of a stature to grasp and to celebrate the philosopher-poet who in the unspecifiability of his thoughts had on more than one occasion made the descent into Chaos and enquired of the elements their origin, law, harmony. . . . It is painful to admit that he fell below the task allotted him." Another such knot centers round Samuel Butler, novelist, satirist, writer on theories of evolution in the 1870's, campaigner on behalf of Lamarck and Erasmus rather than Charles Darwin, whose ideas are taken up by George Bernard Shaw and put together with those of Bergson.

What world are we in here? Certainly not the tidy (and unreal) world in which we mostly live where the disciplines are separated, but the world of what one might call the virtue of science and of poetry as of all other noble disciplines of learning, the central Method of the thinking and divining human intelligence, which is to be worked at and advanced by all who so can and are called to this. It has been, during this hundred years, the work mainly of scientists and poets. (The play and the novel I mentioned earlier which seem to me really to deal with this virtue are Shaw's *In Good King Charles's Golden Days* and Hermann Hesse's *Das Glasperlenspiel*.) Yet as we sketch out the century's curve of this virtue at work, we shall notice an interesting thing. In the 1860's it was the poets who were standing to the task. There comes a turning point about halfway; and now, it seems to me, the scientists have taken over, and it is they who in our own day carry forward the Method and also, I believe, the essential poetic vision.

Of the poets in the 1860's we take, first, Walt Whitman and his "Passage to India." I am puzzled as to how to conjure up this wonderful but longish poem here; stray quotations cannot compass it, and probably all I can do is ask you to pull it up yourselves out of your own memories. It begins with technology and geography, the roundness of the world felt by the poet as he stands on the crest of America and watches, as it were, eastward the opening of the Suez Canal, westward the progress of the transcontinental railroad, and sees in the human communications "the rondure of the world at last accomplished." Then he looks back to the history of this rounding of the world and what it has meant, myths and science woven together, the great sea explorers bringing us not merely the merchandise but all the myths of the East; and he turns our minds back to our biological history, Adam and Eve, and he says, "down from the gardens of Asia descending, then their myriad progeny after them," all humanity being, as the seafarers were, "yearning, curious, with restless explorations." And it does not stop with the rondure of the world, magnificent hymn though the poet sings to that. There is here, he says, challenge to seek passage to primal thought, to more than India, this exploration taking on the image of launching out into space and (it is the same) of launching out into God:

> O thou transcendent, nameless,
> The fibre and the breath,
> Light of the light,
> Shedding forth universes,
> Thou centre of them. . . .

Of the poem as a whole Whitman said, "There's more of me, the essential ultimate me, in that than in any of the poems . . . the burden of it is evolution—the one thing escaping the other—the unfolding of cosmic purposes."[1]

It may be that Hugo was attempting something not altogether different in *La Légende des Siècles,* but it is not that voice I want here but his essay on Shakespeare, where he discusses the nature of genius as shown in the imagination of artist and scientist alike, that very virtue in fact which, I have suggested, is the essential unity between science and art. He has wonderful things to say about science, signalizing not its clarity, logic, control, as so many of his contemporaries did exclusively, but its groping and tentative nature: "Tout ce long tâtonnement, c'est la science. Cuvier se trompait hier, Lagrange avant-hier, Leibniz avant Lagrange, Gassendi avant Leibniz . . . Oh l'admirable merveille que ce monceau fourmillant de rêves engendrant le réel!" He recognizes too the crisscross nature of genius—"Dante combines and calculates, Newton dreams"—and its spring and source in the human imagination. We should remind ourselves here of the great work of instauration of the imagination undertaken from the start of the nineteenth century by poets, Blake, Coleridge, Wordsworth; now Hugo picks up that note and prophesies its injection into science and mathematics. In an extraordinary passage in this essay Hugo goes so far as to claim (further than Sainte-Beuve) that in mathematical operation, in algebra and geometry and calculus alike, all of which he mentions and more, "l'imagination est le coefficient du calcul et les mathématiques deviennent poésie."

It is now only a small step to what I called the midway turning point in the curve or arch of these hundred years. It comes, it seems to me, with Henri Poincaré, the mathematician, and the three works in which he examines the workings of the mathematical and scientific mind, *La Science et l'Hypothèse*, 1902, *Science et Méthode*, 1905, and *La Valeur de la Science,* 1906. What interests him, and what he pursues and insists on, is the function, in mathematical thinking, of that which is not logic—imagination or intuition which, he says, is the instrument of invention or discovery, where logic is the instrument of demonstration (this in Chapter I of the last-named work of his). It is an emphasis destined to be taken up more and more by scientists, especially perhaps the mathematicians and physicists, right down to our own time, ousting slowly but surely the previous notion of science as a clear and entirely rational occupation. Not long ago I made for myself a list of the metaphors or terms which scientists themselves have applied to their own sort of thinking, the thinking which is not just routine but invention. It includes leaps, bridges (over what chasms, one wonders?), guessing, divining, inspiration, seizing on a happy idea, somnambulism, dreams, music, moving nets, bodily metabolisms, the dance, and so forth. If this gives us a sense of vertigo, we can on Hugo's crisscross principle steady ourselves with a poet saying, "I could not bear, from 1892 onwards, that the poetic state should be set in opposition to the full and sustained activity of the intellect," the poet who took *Ostinato rigore* from Leonardo as his device—Paul Valéry again.

It seems to me possible, if we slide down now to the end of our big curve of time, to match the respective visions of a united science and poetry which we saw in Whitman and Hugo with the work of two great scientists contemporary with ourselves. If we take the latter vision first, as part of the advancing enquiry into the nature of thinking and imagination, in science as in poetry, we shall find ourselves now in the company of Michael Polanyi; and I have specially in mind his most recent work, lectures given at Duke University in 1964, under the title *Man in Thought,* which are due for publication. As with a long poem, we cannot summarize adequately; but of special interest in our context here, besides the work he has already done upon the function of tacit knowing, of focal and subsidiary awareness and our powers of attending *from* the tacitly known *to* that which is yet to be known, is his discussion of the inherent unspecifiability of all good thinking (which is to say "fruitful," which is to say "true thinking"—is inexhaustible that which is real, in fact) and hence of all great scientific discovery and the theory that results from it. And if we take the other line from where we began, the stupendous vision in Whitman, this also can be matched and taken forward in our day: in the work of Dr. Joseph Needham, the volumes, already published and yet to come, of *Science and Civilisa-*

tion in China. Dr. Needham is a methodologist of the scientific imagination in his own right (look for it in the passages on the metaphors of science Eastern and Western, and his exposition of "correlative thinking"); but it is the global view one catches here, as in "Passage to India," his pioneering work taking us again round the rondure of the world and opening up for us the scientific imagination not only of China but of India, of Babylon, the great Moslem civilization—they all come in.

It is to the great scientists of our day, I am convinced, that we must look for the central work on the methodology of the human imagination which I have called the virtue of both science and literature, the essential unity. A hundred years ago this work was being done mainly by poets. Now the scientists have it, and the reason for this is that science is where, today, thinking is going on. It is not going on in the arts (practicing) or the Arts (academic). It may seem strange that I who am arts and Arts, as it were, should say so, but I think only somebody thoroughly involved in the Arts now could realize our utter unfamiliarity with thinking as such, an unfamiliarity so complete I doubt whether we should even recognize it if we saw it. Which brings me to the end of my hundred-year arch, and, for a moment or two, to the subject of education.

It may have seemed to you that what I have been saying here about the unity of science and literature is, even if admissible in theory perhaps, impossible in practice. We cannot, we are told, master the whole of one field these days, let alone more than one. The days are gone when men could take all learning for their province, are they not? Or so we will hear Arts people say, who, we suspect, may be unaware of the work and the quality and range of mind of men of science such as D'Arcy Thompson or Dr. Needham or Professor Evelyn Hutchinson. The reason we think this to be impossible is because our education tells us so. That is all. I have at this point reached the stage of my discussion where I have to burst into some sort of personal testimony, for I feel as I imagine St. Paul would have felt if he had been asked to discourse, say, on "The Psychology and Significance of Religious Conversion During the Last Hundred Years." He would have managed quite well up to a certain point, and then he would have had to blow his top and say, "Excuse me, but I *know* something about this, myself, at first-hand." So do I, for I underwent a totally unexpected and spectacular conversion to mathematics and science at the age of twenty-seven, and I have to say something about it. Before I do, however, I want to bring in two bits of evidence in support of what I have to say—of the view, which our education does its best to blind us to, that anyone who can think at all is capable of moving freely in sciences and arts both. The first is R. G. Collingwood, who says in his autobiography, regarding himself as he finished school and headed for the university, "I was equally well fitted to specialize in Greek and Latin, or in modern history and languages (I spoke and read French and German almost as easily as English) or in the natural sciences; and nothing would have afforded my mind its proper nourishment except to study equally all three." As one says amen one reads, depressingly, the next sentence, "I had to specialize in something." The other piece of evidence I draw from my own experience of teaching Negro students in the southern United States, who do not have our advantages in education and so are not properly indoctrinated with the idea that they are either mathematicians or writers or economists or biologists or painters. I can think straight away of four students of mine in recent years with an extraordinary range of talent right across the board. I do not think this is a peculiarity of the Negro race; I think it is one of the strange advantages conferred by a bad education.

All this in parenthesis, in the space between the end of the big hundred-year arch we are concerned with, and the little twenty-year one which is my own experience and which we will skim over fairly quickly, marking our progress with three unpublished poems of my own which belong in the story.

I was, as a child and right from the start, a word-child, not a number-child, and always wanted to write, so that already by the age of eleven I was clear, as was everybody else round me, that I was an Arts and Humanities type. I had to endure the imposed boredom of school mathematics and of minimal school science until my school-leaving certificate, in which I just got through the mathematics requirements and vowed never to be troubled with that dreary subject again. I read Modern Languages at a university which does not even require a "Minor"—you just do your own single subject and that is all. After three years absence during the war I returned to do graduate work, my subject being late nineteenth-century French poetry. Finding that I needed to think about the nature of language and poetry in general and that a vast number of works existed already on this subject, I took, in October 1946, the half-playful decision to pretend that no one had ever thought about language and poetry before and to begin to think about them myself. The result was unexpected and appalling. I found I didn't know how to think. That is understatement—I hadn't the faintest idea what thinking was or how one began on it. I was in my fifth year of formal university training.

In the sudden eclipse and darkness which descended on me I tried to grope my way towards some discovery of what thinking was. My preliminary notions were that it seemed to have something to do with watching what went on in one's own mind and with relations between things. I found one could get no grip on things themselves; but the relations between them, and the movements of one's mind, offered possibilities. Of relations I knew nothing. So ignorant was I that I thought I had better go and read about relativity because it sounded as if it might help. It was not the right answer, but was not too far wrong. One day, as I sat thinking about words and scribbling down notes on a paper in front of me, I found I had made a note about words being apprehensible, perhaps, as "complex variables." That sounded to me like a term in higher mathematics. By now I was prepared to follow any clues, so I bicycled down to college, went into the section of the library labeled Mathematics, which I had never before entered, and looked around, with a strong sense of comedy, asking myself where I should begin. At eye-level opposite I caught sight of something called *Principia Mathematica* by Whitehead and Russell. It sounded promising. I pulled out Volume I and began at the beginning. Naturally I could not understand three-fifths of what I read. But the remainder was a revelation. I had found what I was looking for. Here were the people who knew about relations in the mind, and they sent me on to the symbolic logicians and the physicists. I read almost nothing but mathematics and physics for a year, with a passion of delight, and much remonstrance, later, from those in authority.

The first poem belongs here. It is rather a childish one, as befits my situation, but it does express the sense of utter amazement and joy at finding that what I had supposed to be an alien if not hostile world to the poet was nothing of the sort. I called it "On First Reading Some Modern Physicists." (The image, by the way, I found in Sir James Jeans, who says that in size man comes about midway between a nebula and an atom.)

>Between the atoms and the nebulae
>   I stand,
>With an infinity
>   On either hand,
>
>Holding by blessed chance
>   The golden mean,
>Delicately to dance,
>   Between, between.

> Secure from any sense,
>   Nonsense divine,
> Dear double influence,
>   Dear friends of mine.
>
> At peace with all that is,
>   I softly see,
> To left and right, bright atomies,
>   Bright nebulae.

The second poem speaks of a very different reaction, for I should not be honest if I left you with the impression that this sort of conversion brings nothing but joy. The second stage is terror—at the collapse of one's whole universe of intellect, at the helplessness and the darkness, at the unfamiliarity, beautiful yet seemingly threatening too, of the ways of thought in mathematics and science. Not that there is anything amiss with fear as a discipline: the twin disciplines of joy and fear, as Wordsworth testifies. I only give one stanza of this poem. There were three, but the first and third are unsatisfactory, and this says enough.

> Star-needles through the temples and the eyes
>   Draw tight the thread.
> Useless henceforth for any enterprise
>   This piecework head,
> Except to contemplate growing complexity
>   With growing dread.

For the third poem I could have taken almost any of those that make up Part V of *The Orphic Voice,* they and the book carrying forward this theme of the two threads of science and poetry which are separate and yet love each other so deeply, indeed, are wedded one to the other. But instead I prefer to end my twenty years' span and this discussion with a much more recent poem, written only two years ago on the same theme, for it goes on and on—"Is inexhaustible that which is real." Like the previous pair, it is also about the stars, but this time the two voices are heard as a dialogue, science and poetry, man and woman, both voices within ourselves and every thinking mind, you understand. For in the end one wants to speak simply by images and love, and that is why I finish with this poem. I have not tried to publish it before. It seems as if it had been waiting for its own time and place, and now finds it here.

Dialogue

He. Speak first of what you see.
   There is a change I sense
   Upon the instruments
   That scan these mortal skies.

She. Phenomenon it seems
   Among the heavenly host;
   Strangely I see it most
   Behind closed eyes.
   Golden . . . a cosmic cloud
   Comes drifting over space,
   A whorl of active stars
   Has drawn it to a close;
   Through that suffusing veil
   Bright silver points career.

He. Two systems wedded here.

She. Rather, a single rose.

He. Soft-shining vaporous globe
   Darted by busy sparks
   The telescope confirms
   On the blue curve of night,
   Dialysis of light,
   Conjunctured galaxy.

She. Or else a life that hangs
   In absolute jeopardy.

He. Interpret that we saw.

She. You know not what you ask.
   Poor tapers as we are
   In clouds of clay,
   How can we hope to win
   The meaning of a sign?

He. Beloved mine,
   Breathe deeply, and begin.

She. Stars then—o Hyblean bees,
   Most marvellous of swarms,
   That breed in the abyss
   The mind's incipient forms,
   Yet influence too fierce
   Unless in figures sung—
   Light . . . sweet Danae . . . one kiss
   Of honey on the tongue.

NOTES

1. Horace Traubel, *With Walt Whitman in Camden* (1915) I, 156–157; quoted in Emory Holloway, ed., *Walt Whitman: Complete Poetry and Selected Prose and Letters* (London, 1938), 1076–1077.

2

SCIENTIFIC AND PHILOSOPHIC REVOLUTIONS

Richard McKeon

Accounts of recent developments in science and philosophy frequently use the word "revolution." The two revolutions are usually related to each other and to a revolution in the circumstances and conditions—political, economic, social, and cultural—which underlie and influence scientific and philosophic thought and which in turn are modified and formed by scientific advances and philosophic ideologies. Since the revolutions in science, philosophy, and society influence one another reflexively, it is difficult to state the relations between science and philosophy unambiguously. A historical retrospect is an examination of past facts in their bearings on the present situation. The past facts in the development of science and philosophy, however, are problems encountered and theories formed and tested: the facts of history are selected and established by ideas, and the facts of intellectual history, like those of political, social, or economic history, express more or less explicitly a philosophy of history. The relation between science and philosophy during the last one hundred years has been traced in two broadly different ways: the revolutions in science resulted from the examination of new data and the reformulation of basic concepts to accommodate them, while the revolutions in philosophy resulted from the need to restate philosophic problems to accommodate science and common sense; or the revolutions in science were philosophic revolutions in the interpretation of phenomena and in the formulation of laws to order them, while the revolutions in philosophy were refinements of problems of experience common to science and philosophy and the application of the same or like methods to the solution of philosophic problems. The differences between the two accounts is not simply a question of "facts": questions of the relation of science to philosophy, and even of the history of science to the history of philosophy, are at bottom philosophic questions.

The two historical accounts differ in the facts they relate as well as in the philosophies they express. According to the one mode of treating them, the revolutions in science began to take form a hundred years ago when Clerk Maxwell and his contemporaries were led by examination of phenomena of fluids, gases, heat, light, electricity, and magnetism to formulate laws of dynamics which went beyond Newtonian mechanical laws of mass-particles, reinforced by like fundamental changes in biology, psychology, and mathematics, while the revolutions in philosophy began to take form at the beginning of the twentieth century, a little more than fifty years ago, in the innovations of pragmatism, realism, linguistic philosophy, phenomenology, and existentialism. According to the other mode of treating them, the revolutions in science and philosophy both began to take form a hundred years ago in the recognition and treatment of new problems, and both moved about fifty years ago to new modes of formulating theories and explanations. In the first account, the history of the revolution in science tends to be told as a history of progress in knowledge and in the discovery of facts and the establishment of truths, while the history of the revolution in philosophy tends to be told as a history of the abandonment of meaningless statements, unreal problems, and unwarrantable assumptions. In the second account, the history of the revolution in science and philosophy tends to be told as a history of the transition from a search for methods of proceeding from the known to the unknown and of establishing basic and comprehensive laws and theories to a search for explanations of facts and for resolutions of concrete problems by the establishment of lawlike or statistical interpretations.

The difference between the two accounts is not one to be resolved by determining what "in fact" occurred, because it is a difference in the philosophic interpretation of the relation between "fact" and "theory": the first is an account of the accomplishments of scientists and of the statements of philosophers as "facts," and the second is an account of the "problems" treated by scientists and philosophers, of the similarities and differences of their methods, and of the stages of transition in their modes of stating problems and their methods of resolving them. The difference between the two accounts reflects different conceptions of the relation between philosophy and science which extend back beyond the beginnings of modern science to problems of the relation of philosophy to *scientia* and to *episteme* (and includes the problem of the relation of "science" to "scientia" and "episteme"): from the beginnings of philosophy some philosophers have conceived their task as the application of the methods of "science" to philosophic problems, and they have treated the history of philosophy in the first mode as an account of efforts in that venture; and from the beginning other philosophers have conceived their task as the refinement of the methods of knowing and philosophy as a culmination of the "sciences," and they have treated the history of philosophy in the second mode as an account of the extension and modification of assumptions and principles. Philosophy is or becomes a science in the first account; philosophy is distinct from science, or science becomes philosophic, in the second account.

The distinction between the two modes of intellectual history, since it is based on a philosophic distinction between facts (including the facts recorded concerning the formation and statement of theories) and theories (including the theories used in the determination of facts), is a distinction between modes of philosophizing as well as between modes of recording the history of philosophizing. The history of philosophy may be viewed as a record of facts, that is, of the positions taken by philosophers; and philosophizing may be treated as controversy and discussion of oppositions between the positions. The history of philosophy may also be viewed as a record of problems, that is, of the issues discussed by philosophers; and philosophizing may be treated as controversy and discussion of the implications and consequences of different resolutions of common problems. Statements of facts are subject to discussion and controversy to determine their meaning and to establish their truth or falsity; statements of problems are subject to discussion and controversy to explore the range of their meanings and the variety of facts to which they apply. Oppositions are found and explored in discussion of *facts* only if the statement of fact is or can be made unambiguous in noncontradictory statements of definition, inference, and application. Oppositions arise and lead to divergent developments in discussion of *problems* only if the statement of the problem contains a productive ambiguity, that is, an open possibility of more than one tenable resolution, which is the source of suggestive inconsistencies, that is, a hypothetical entertainment of assumptions prior to examining their relation to other principles already assumed.

The positions taken by philosophers and scientists in any period may be stated in a schema of opposed mutually exclusive positions or in a nexus of interconnected ambiguous questions. If the examination of the retrospect of a hundred years of science and philosophy is for the purpose of securing insight into the next hundred years, the history of problems treated has an advantage over the history of positions once held or now thought to be warranted. In the latter mode of history science begins with an initial specious superiority over philosophy (the history of science seems to be cumulative, the history of philosophy noncumulative); in the former mode of history, since many of the controversies of philosophy and even of science are seen to be answers to different questions rather than simple oppositions, the ambiguity of common problems may be made the basis for the determination of precise issues and the ambiguity of the terms in which they are stated may be made the basis for the discovery of the same issues in different contexts and in different terminologies.

An account of the problems and issues of the last century may be begun at any point of the dense series of controversies which contributed to advancement of knowledge of facts and to change of methods in formulating questions and establishing solutions. In a retrospect of one hundred years, the date 1865 is a neutral arbitrary beginning point, and since John Stuart Mill was involved in numerous controversies during that year, the positions opposed in his controversies are an entrance point into the problems of science and philosophy during the last century. In 1865 Mill published two books, *An Examination of Sir William Hamilton's Philosophy and of the Principal Philosophical Questions Discussed in his Writings* and *Auguste Comte and Positivism*. Both are unusually detailed and fair discussions of philosophical questions, the one to establish the grounds for refutation of the *a priorism* which Hamilton derived from Kant and the Scottish Common Sense philosophy, and the other for controversy with positivism. Five years earlier Mill's running controversy with Whewell concerning induction, which he had conducted in the text and footnotes of successive editions of his *System of Logic,* had come to a high point in Whewell's expansion and republication of his essay, "Of Induction, with Special Reference to Mr. J. Mill's System of Logic," in *On the Philosophy of Discovery*. Mill presents the pattern of these controversies in terms of method: "The philosophy of Science consists of two parts; the methods of investigation, and the requisites of proof. The one points out the roads by which the human intellect arrives at conclusions, the other the mode of testing their evidence. The former if complete would be an Organon of Discovery, the latter of Proof."[1] Hamilton neglected induction; Comte took no account of proof or the methods of testing the laws of phenomena; Whewell confused induction with discovery. The philosophical examination of these issues in the years since Mill stated them has been an exploration of the interrelations of methods and structures which emerge from the ambiguities of discovery and proof, induction and deduction, analysis and synthesis. The scientific use of the methods described by philosophers and scientists in that period has been an exploration of problems encountered and of facts discovered.

Mill's choice of issues was determined by the state of science in his times. He records that when he was working on his *Logic* in 1837, he found his materials in Whewell's *History of the Inductive Sciences,* which had been published that year, but he was guided in his thoughts by Herschel's *Discourse on the Study of Natural Philosophy*.[2] He had completed his study of induction when he read the first two volumes of Comte's *Cours de Philosophie Positive,* which had been published in 1830, and he notes Comte's precise and profound statement of the method of investigation but his omission of any definition of the conditions of proof. Herschel's *Discourse* had also appeared in 1830, a year after the publication of Charles Babbage's "Decline of the State of Science in England" and the foundation of the British Association for the Advancement of Science. Herschel analyzes the growth and decline of science after Newton: "The immediate successors of Newton found full occupation in verifying his discoveries, and in extending and improving the mathematical methods which were to prove the keys to an inexhaustible treasure of knowledge." Newton's countrymen, however, abandoned the task of "mathematico-physical discovery" to German and French inquirers, to Clairault, D'Alembert, Euler, Lagrange, and Laplace. The simultaneous discovery by Leibniz of a method of mathematical investigation "in every respect similar to that of Newton" stimulated Continental geometers to cultivate it and to impress on it "a character more entirely independent of the ancient geometry, to which Newton was peculiarly attached."[3]

Herschel's statement of the tasks undertaken by Newton's successors uncovers issues in the meaning and use of the methods of "mathematico-physical discovery." The work of Continental mathematicians and astronomers on Newton's discoveries and mathematical methods had built on differences between Descartes' mathematical method and Newton's mechanical method and had made "mathematico-physical discovery" mathematical rather than physical.[4] Newton had sought a "universal mechanics," since geometry and all other sciences are founded on mechanical practice. Descartes had sought a "universal mathesis," since mechanics and all other sciences follow the order of mathematical demonstration. Both mathematico-physical methods were applied to the task of explaining all physical nature in terms of matter in motion. The methods were the same or were subject to translation or transformation one into the other, yet even in the case of identical methods, as Herschel suggests in the case of the calculus, they were oriented to different problems and they treated different facts. The issues of problem and the issues of fact were subject to controversy in scientific inquiry, and the resolution of problems step by step opened up new problems and encountered new paradoxes. The differences in the scientific methods were made the basis of the philosophic opposition between what has been called "rationalism" and "empiricism," and the development of the opposed positions has been a history of doctrines posited in fact rather than of problems investigated in inquiry. The difference between the history of science and the history of philosophy, as they tend to be presented, is due less to a difference between science and philosophy in fact than to a tendency to subordinate oppositions of doctrine to discussion of problems in the one history and to subordinate statement of problems to doctrines and their oppositions in the other. The Cartesian universal mathematics was not "rational" nor was the Newtonian universal mechanics "empirical"; they were both mathematico-physical methods to explain particular phenomena and to establish universal laws, and they differed concerning the mathematico-physical entities and mathematico-physical wholes relevant to those tasks.

The mechanical method was used for the statement of laws of motion and their application to all phenomena. In 1883 Ernst Mach found no diminution in the two merits of Newton: "The merits of Newton with respect to our subject were twofold. First, he greatly extended the range of mechanical physics by his discovery of *universal gravitation*. Second, he *completed the formal enunciation of the mechanical principles now generally accepted*. Since his time no essentially new principle has been stated."[5] The work of Lagrange and Laplace contributed to this view of the Newtonian mechanics. Lagrange extended and improved the mathematical methods in his *Mécanique Analytique* (1788) and his *Théorie des Fonctions Analytiques* (1799), and Laplace verified and systematized its applications in his *Exposition du Systéme du Monde* (1796) and in the five volumes of his *Mécanique Celeste* (1799–1825), which was so closely tied to the Newtonian physics that it came to be called the second edition of the *Principia*. For some thirty years doubts concerning the sufficiency of the doctrine of universal gravitation to explain all cosmical phenomena were dispelled by Laplace's work. In the meantime the mechanical method was extended to the phenomena of heat, light, and electromagnetism. Imponderable fluids and inert corpuscles were added to ponderable masses, but the phenomena of electricity and magnetism led to the construction from Faraday's notion of a field of the concept of continuous fields, which are not mechanically explicable, and its substitution for the concept of continuously diffused masses. Clerk Maxwell makes use of what he calls dynamical explanation, that is, the description of a physical phenomenon as a change in the configuration of a material system, in his *Electricity and Magnetism* in 1873. The issue of universal mechanics and universal mathematics was still present in his formulation: Maxwell expounds his conception of the dynamic method in contrast to Lagrange's method of reducing the ordinary dynamical equations of motion of the parts of a connected system to a number equal to the degrees of freedom of the system. The aim of Lagrange, he says, is to bring dynamics under the power of the calculus; his aim is to cultivate our dynamical ideas, and he undertakes to retranslate the results of the labors of the mathematicians from the language of the calculus to the language of dynamics.[6] The emphasis is shifted from the material point to the continuous field, from the mass point to energy. According to Maxwell, a complete discussion of energy would include the whole of the physical sciences, and the principle of the conservation of energy guides inquiry in all phenomena. But paradoxes concerning masspoints and the ether as the medium of the phenomena of light and electromagnetism were to inspire a major revolution in mechanical or dynamical method.

The mathematical method was used for the discovery of principles and the demonstration of conclusions. *Mathesis universalis* includes all sciences, since it treats of order and measure without reference to any particular matter.[7] Descartes argued that no other principles are needed in physics than in geometry or abstract mathematics, "because all the phenomena of nature can be explained by their means, and sure demonstrations can be given concerning them."[8] He sought a science of necessary demonstrations and a true method to discover all truths methodically. He distinguished between order (*ordo*) and manner of demonstration (*ratio demonstrandi*) in the geometric mode of writing. Order is the sequence from things posited and known first to what follows and is demonstrated from them. There are two manners of demonstrating: analysis is the true way of discovering things methodically and *a priori* (that is, by showing how effects depend on causes); synthesis is the opposite way of demonstrating clearly that which is concluded *a posteriori* (that is, by examining causes through their effects), and it uses long series of definitions, postulates, axioms, theorems, and problems.[9] The three treatises appended to the *Discours de la Methode pour bien Conduire sa Raison et Chercher la Verité dans les Sciences*—the *Dioptrics, Meteors,* and *Geometry*—illustrate the use of the analytic method. The principles of physics are mathematical and make no use of Newtonian material points or absolute space, time, or motion; they were criticized from the point of view of the universal mechanics for committing the error of confusing space with matter. Analysis is a method of discovery and proof; later Cartesians sometimes used the synthetic method of Euclidean geometry as a method of discovery and proof. Leibniz argued that Descartes had not reduced the synthesis of geometers to analysis, since his analysis was an analysis of numbers, not of lines, and he succeeded in reducing geometry only indirectly, since magnitudes can be expressed by numbers.[10]

Mathematical principles and physical principles are identical in mathematico-physical methods. The identity is based on assumptions about the nature of motion and of moving objects in the universal mechanics and about the nature of order and of instances of order in the universal mathematics. Galileo based the identity of mathematical and physical principles on the real existence of indivisibles and infinitesimals, Newton on mass-points and centripetal force, Vico on the fact that mathematical demonstrations and truths in physics are both made. Descartes based it on order and measure, Leibniz on the rationality of the real. The universality of the universal mechanics consisted in its application to all motions in the system of the world, but during the nineteenth century the extension of mechanics to optics and electrodynamics led to the substitution of fields for forces as fundamental variables. The universality of the universal mathematics consisted in providing a method for treating all problems, but during the nineteenth century differences of method in mathematics led to the substitution of varieties of principles used synthetically for the universal principles applied analytically for the discovery of all truths. The methods of maxima and minima, of analytical geometry, analysis situs, infinitesimal analysis, combinatorial analysis, which were universal in scope, were diversified in non-Euclidean geometries, projective geometry, axiomatics, non-Archimedean geometries, sets, and the fundamentals of arithmetic. The two methods influenced each other as they underwent converse revolutions from indivisible to fields and from continua to sets, and their merged influence was apparent in the two revolutions of the early twentieth century in which relativity physics transformed the concepts of space and time in general field equations and quantum mechanics transformed the conceptions of motion and position in probability laws.

The development of philosophy during the same period can be stated in terms of the same issues and problems. The issues derived from Mill's controversies a hundred years ago isolate the problems and methods of science. Since they were stated controversially, they tend to determine the oppositions and refutations of philosophers and to support the impression that science differs from philosophy, but they can also be used to isolate the problems and methods of philosophy. The resulting matrix of philosophic principles and methods yields distinctions and connections which are continuous with those employed in scientific inquiry, and it opens up interpretations of the philosophic revolutions of the early twentieth century in terms of the emergence of problems and the interpretation of statements rather than the enunciation of positions held and the refutation of positions opposed.

The structure which Mill gave to his controversies is determined by his conception of the parts of logic and the relation of logic to scientific method. The distinction of the two determinants and their two subdivisions is expressed in the title of his logic—*A System of Logic, Ratiocinative and Inductive: Being a Connected View of the Principles of Evidence and the Methods of Scientific Investigation*. The two parts of logic reflect the two processes of science—proof and discovery, evidence for conclusions and investigation of problems. In the tradition of Newton and the mechanical method, the two parts of science are inseparably connected —laws are constructs discovered and proved. Mill's controversy with Whewell was in this tradition. Induction establishes a connection between occurrences, and Whewell's error, according to Mill, was to seek the cause of the connection in ideas and to confuse induction with invention. To separate the two parts of science is to fall into one of the two opposite errors of concentrating on deduction or on investigation of phenomena, which led to bifurcations in the tradition of Descartes when a succession of syntheses took the place of mathematical analysis. Hamilton defined logic as "the science of the laws of thought" and reduced logic to deduction and the syllogism apart from a theory of induction. He denied the possibility of a Philosophy of Evidence and of the Investigation of Nature.[11] Comte, on the other hand, was incomparable in his treatment of methods of investigating phenomena, but he supplied no test of proof and he saw no use for deductive logic.

The matrix which defines the semantic oppositions of meanings and doctrines may be used to differentiate the forms and methods of inquiry in philosophy and science and to state the problems treated in the late nineteenth and early twentieth century in a continuing inquiry concerning method and science. It was a period in which theories of the nature of science and of scientific law and organizations and classifications of the sciences were inseparable parts of the investigation of the methods of sciences and the elaboration of the methods of logics. The issues presented by different conceptions of science and different sciences were the source and the subject matter of the different methods constructed, modified, and extended. The common question underlying the issues treated from the perspective of different philosophic assumptions was, What is the structure of science: on what is it based and by what method? Consideration of the positions taken yields a history of philosophical oppositions defined with semantic rigidity; consideration of the problems treated yields a history of philosophical inquiry following through the hypotheses formed to define ambiguous common questions. The common basis of Mill and Whewell was their agreement that the structure of science was found in tested constructs based on facts and ideas, and their differences concerning whether facts or ideas were fundamental led them to form different concepts of facts and ideas and different theories of induction and deduction, discovery and proof. The common basis of Hamilton and Comte was their agreement that the structure of science was found in regularities encountered in the processes of thought or of nature, and their differences concerning whether forms of thought or forms of occurrences were fundamental led them to form different concepts of synthesis, *a priori* and positivistic, and different theories of the relation of pure logic to modified or concrete logic and of subjective to objective methods and syntheses.

Mill reduced all proof and discovery to induction and the interpretation of inductions. "We have found that all Inference, consequently all Proof, and all discovery of truths not self-evident, consists of inductions, and the interpretation of inductions: that all our knowledge, not intuitive, comes to us exclusively from that source."[12] The emphasis is on the logic of induction because Mill designed the System of Logic to supply, in opposition to the *a priori* view of human knowledge, a doctrine "which derives all knowledge from experience, and all moral and intellectual qualities principally from the directions given to the associations."[13] There are deductive or ratiocinative sciences as well as experimental sciences, but the first principles of all sciences, including geometry, are results of induction.[14] Every branch of natural philosophy was originally experimental, and although "mechanics, hydrostatics, optics, acoustics, thermology have successively been rendered mathematical and astronomy was brought by Newton within the laws of general mechanics . . . every step of the deduction is still an induction. The opposition is not between the terms deductive and inductive, but between deductive and experimental."[15] Mill defines induction as "generalization from experience" or "the operation of discovering and proving general propositions."[16] The deductive method consists of three operations: direct induction, ratiocination, and verification.[17] There are only two ways in which laws of nature can be ascertained—deductively and experimentally.[18]

Mill goes to the sciences for the explanation and analysis of these methods. Mechanics is a deductive or demonstrative science and chemistry is not, because the same law applies in mechanics to the composition of cause, to the effect of each cause acting separately, and to the effect of the two together, whereas in chemistry it does not, the taste of sugar of lead is not the sum of the tastes of its component elements. This is even more true of organized bodies than of chemical combination.[19] Chemistry uses the experimental method, and the four experimental methods governed by the five canons of experimental inquiry are methods of *chemical* inquiry. They are instruments of discovery and proof of laws, sequences, and effects. "To ascertain, therefore, what are the laws of causation which exist in nature; to determine the effect of every cause, and the causes of all effects, is the main business of Induction; and to point out how this is done is the chief object of Inductive Logic."[20] The deductive method is divided into the *geometrical* or abstract and the *physical* or concrete. The experimental or chemical and the abstract or geometrical modes of investigation are not properly used in social science, but the concrete deductive or physical and the inverse deductive or historical methods are the methods proper to inquiry into social phenomena.[21]

Whewell likewise treated induction as the basic process in science and method, and his distinction between facts and ideas has an ambiguous congruence with Mill's distinction between discovery and interpretation of the laws of phenomena. In their divergent interpretations of the ambiguous common questions they treated Mill sought to examine induction as proof, Whewell as discovery. *The History of the Inductive Sciences* and *The Philosophy of the Inductive Sciences*, Whewell said, were intended to show the steps of development of the sciences and the philosophic principles involved in those steps.[22] *The Philosophy of the Inductive Sciences* has two parts, the *History of Scientific Ideas* and the *Novum Organon Renovatum*. The two histories differed in that the former presents the history of the sciences so far as it depends on observed facts, while the latter treats the ideas by which such facts are bound into theories.[23] Since each step of the history was a scientific *discovery* in which a new conception was applied in order to bind together observed facts, Whewell expanded some of the chapters of *The Philosophy of the Inductive Sciences* and added views of his own, under the title *Philosophy of Discovery,* because it was the novelty of each step, not the fact that the conjunction of observations was in each case a logical induction, which marked the advance of the science, and the philosophy he sought was therefore not the philosophy of induction but the philosophy of discovery.[24] The method which he sets forth in the *Novum Organon Renovatum* makes use of the same distinction: there are two principal processes by which science is constructed, the Explication of Concepts and the Colligation of Facts.[25] In a strict sense an art of discovery is not possible, but the process of discovery can be resolved into its parts, and rules and methods can be given for each portion of the process.[26] The process by which science is constructed has three parts: the Decomposition and Observation of Complex Facts; the Explication of our Ideal Conceptions; and the Colligation of Elementary Facts by means of those Conceptions. No methods can be given for the explication of conceptions, although something may be done by education and discussion of ideas; the methods of exact and systematic observation are largely processes of measurement, dependent on the ideas of number, space, and time; the methods of induction are methods for the discovery of laws of phenomena and causes, and the process of induction may be resolved into three steps: the selection of the Idea, the Construction of the Conception, and the Determination of the Magnitudes.[27]

The classification of the sciences does not depend "upon the faculties of the mind to which the separate parts of our knowledge owe their origin, nor upon the objects which each science contemplates; but upon a more natural and fundamental element—namely, the Ideas which each science involves."[28] The resulting classification is a hierarchy in which the prior ideas may continue to be used in the posterior sciences: pure mathematical sciences (geometry, arithmetic, algebra, differentials—ideas of space, time, number, sign, limit), pure motional sciences (pure mechanism, formal astronomy—idea of motion), mechanical sciences (statics, dynamics, hydrostatics, hydrodynamics, physical astronomy—ideas of cause, force, matter, inertia, fluid pressure), physics—secondary mechanical sciences (acoustics, formal optics, physical optics, thermotics, atomology—ideas of outness, medium of sensation, intensity of qualities, scales of qualities), physics—analytico-mechanical sciences (electricity, magnetism, galvanism—idea of polarity) analytical science (chemistry—ideas of element-composition, chemical affinity, substance-atoms), analytico-classificatory sciences (crystallography, systematic mineralogy—ideas of symmetry, likeness), classificatory sciences (systematic botany, systematic zoology, comparative anatomy—ideas of degrees of likeness, natural affinity-vital powers), organical sciences (biology—ideas of assimilation, irritability, organization), metaphysics (psychology—ideas of final cause, instinct, emotion, thought), palaetiological sciences (geology, distribution of plants and animals, glossology or comparative philology, ethnography, natural theology—ideas of historical causation, final causes). Induction of laws of phenomena is ordered by induction of causes of phenomena, which depend on the fundamental ideas of substance, force, polarity, and ulterior causes.[29]

For both Mill and Whewell the problems of the structure of science were problems of method, for the structure of science was determined in the construction of science. Induction or discovery is the basic method in constructing science, but "induction" and "discovery" are different methods when "causes" are differentiated into kinds dependent on different ideas connecting facts than when they are treated in common as antecedents of phenomena. In the sense given to induction and deduction in Mill's version of the construction of science, Hamilton had omitted induction and Comte had omitted deduction. Yet both had theories of "induction" and "deduction." Histories of philosophy which are accounts of controversies explore the semantics of one theory in interpretation and refutation of opposed theories, but the controversies provide also the materials for an account of the inquiry concerning method and of the problems treated in the alternative hypotheses based on the different definite meanings given to common ambiguous questions. Hamilton found the structure of science in the regularities encountered in applying laws of thought; Comte found the structure of science in the regularities encountered in hierarchies of phenomena; both found a single method in the sciences and both distinguished general or abstract and applied or concrete sciences.

Hamilton argued that logic is not an organon or instrument, nor is it an art of discovery; it is the science of the necessary forms of thought. He divided logic into general logic (which treats the formal laws of thought without respect to any particular matter) and special logic (which treats these laws in relation to a certain matter, and in subordination to the end of some determinate science).[30] General logic, in turn, is divided into pure or abstract logic (which considers the laws of thought proper, as contained *a priori* in the nature of pure intelligence itself) and modified or concrete logic (which exhibits these laws as modified in their actual applications by certain general circumstances, external and internal, contingent in themselves, but by which human thought is always more or less influenced in its manifestations).[31] Modified logic is not properly an essential part of logic, but a mixture of logic and psychology.

Hamilton follows Kant in dividing logic (pure and modified) into a doctrine of elements or Stoicheiology and a study of methods of Methodology. In pure logic the elements are the fundamental laws of thought (which are four, the principles of identity, contradiction, excluded middle, and reason and consequent) and the products of thought (which are concepts, judgments, and reasonings, that is, the three traditional parts of formal logic, terms, propositions, and syllogisms). In pure logic, methodology consists in method in general (two processes: analysis, which begins with the individual or determined and proceeds to universals and principles, and synthesis, which begins with principles or universals and proceeds to the determined or the individual) and special or logical method (concerned with the form as distinct from the matter of thought, as treated in definition, division, and probation). There are three parts of modified or concrete logic: two concerned with elements—the doctrine of truth and error and the detailed treatment of truth and error; and one with methodology—the acquisition and perfecting of knowledge, which has three parts: experience, speculation, and communication of knowledge. Induction is of two kinds, logical or formal and real or material. Logical induction, like deduction, can be stated in syllogisms; the canons of deductive and inductive syllogisms are different, but they are equally formal: the rule of the deductive syllogism depends on the relation of the containing whole to the contained parts, that of the inductive syllogism on the relation of the constituent parts to the constituted whole.[32] Real induction is part of the modified methodology of experience and observation, and it rests on the constancy or uniformity of nature and on the instinctive expectation we have of this stability.[33]

Truths are distinguished into empirical or *a posteriori* (which have their source in external or internal perception and their form from Understanding or the faculty of Relations, *dianoia*) and pure or *a priori* truths (which are the necessary and universal cognitions by the Regulative Faculty—Intellect Proper or Common Sense, *nous*).[34] The Kantian *a priori* is joined to Scottish Common Sense in applied logic or the doctrine of truth. "The doctrine which has been called *The Philosophy of Common Sense* is the doctrine which founds all our knowledge on belief."[35] It was recognized by Aristotle when he said, "What appears to all men, that we affirm to be, and he who rejects this belief (*pistis*) will assuredly advance nothing better worthy of credit." The classification of the sciences follows from the distinction of Formal Truth from Real Truth. Formal knowledge includes the science of logic and the science of mathematics. The real sciences are sciences of fact, which are of two kinds: mental sciences (which rest on the facts of self-consciousness or the facts of mind) and material sciences (which rest on the presentations of sensitive perception or the facts of nature). The real truths of real knowledge include metaphysical, psychological, physical, and moral or ethical truths.[36]

Comte went to phenomena rather than to thought for the structure of science, but since thought, emotion, and action are also phenomena, education, morality, and religion for the improvement of man make use of a subjective method based on the hierarchy of the sciences established by the objective method. The fundamental law of the development of human intelligence establishes the stages in which extrinsic intrusions of thought are eliminated from the statement of natural law. All our principal conceptions and all branches of our sciences pass successively through three different theoretic stages: the theological or fictive, the metaphysical or abstract, and the scientific or positive.[37] The human spirit recognizes the impossibility of obtaining absolute notions only in the last stage and only then gives up inquiry into the origin and destiny of the universe and into the causes of phenomena. The positivistic method is the method of combining observation and reasoning for the discovery of the effective laws of phenomena, that is, their invariable relations of succession and simultaneity.[38] All phenomena are subject to invariable natural laws; the positivistic method is formed to the precise discovery of those laws and for their reduction to the least possible number.[39] The logical laws of the human spirit can be discovered only by examining the activity of our intellectual faculties;[40] the method of the sciences can be examined only in the sciences in which it is employed.

All sciences, physical and social, are branches of one science and are investigated by one and the same method. Comte, like Hamilton, distinguished two processes in method, analysis and synthesis. Objective analysis yields a classification of the sciences on the basis of phenomena, including the phenomena of human activity; subjective synthesis yields a plan of education and action based on objective analysis for the improvement and use of the sciences, including the sciences of sociology and morals. In the *Course of Positivistic Philosophy* (1830–1842) six basic sciences are distinguished according to natural categories, arranged in an order in which the rational study of each category is based on knowledge of the principal laws of the preceding category.[41] The simplest phenomena are necessarily also the most general phenomena. The initial division of phenomena is into two principal classes, inorganic bodies and organic bodies. The subdivision of these yields five sciences: celestial physics or astronomy; terrestrial physics, subdivided into physics and chemistry; organic physics or physiology; and social physics. The first treats the most general, the simplest, and the most abstract phenomena, which are furthest removed from humanity; the last, on the contrary, treats the most particular, the most complicated, and the most concrete phenomena, which are the most directly important to man.[42] Mathematics is placed at the head of these five, since geometry and mechanics must be viewed as true natural sciences, based as are the rest on observation, although they are capable, because of the extreme simplicity of their phenomena, of an infinitely more perfect degree of systematization which sometimes causes the experimental character of their first principles to be ignored.[43]

True philosophy systematizes, as much as possible, all human existence, individual and more especially collective, and in all three orders of phenomena which characterize existence: thoughts, sentiments, and actions.[44] There is a speculative synthesis, an affective, and an active synthesis, which together may form a grand synthesis. They conform to an exterior objective structure in which the objective synthesis produces sciences and the subjective synthesis a philosophic moral education for human improvement.[45] For the formation of natural philosophy, systematic or spontaneous, the objective method, which proceeds from without to within, from the world to life, is the only one possible. The inverse or subjective method, which proceeds from within to without, from life to the world, contributes to the normal state of our intelligence.[46] The systematization which emanates spontaneously from affective life is theological, and the subjective method must, as does the objective method, give up the vain search for causes and must turn directly to the discovery of laws to ameliorate our condition and our nature, that is, it must cease to be theological and become sociological.[47] The objective analysis provides the necessary basis for the subjective synthesis which condenses all doctrines in morals. Morals, or study of our nature in order to regulate our existence,[48] is added to the list of sciences, and the subjective synthesis, which is single, may operate on seven distinct analytical classifications of the sciences: two binary, two ternary, two quaternary, and finally one quinary—mathematics, physics, biology, sociology, and morals.[49]

Mill's judgment that Comte used the inductive method but had no theory of deduction and made no effort at proof is doubtless his interpretation of Comte's definition of natural law as succession or simultaneity in phenomena and his denial of causes. Inquiry concerning method, in contrast to controversy, must account for Comte's conceptions of induction and deduction by examining Comte's theory of logic and of its relation to science and action. Logic has a place in the objective synthesis, relative to science, and also in the subjective synthesis, relative to action. Comte distinguishes induction and deduction in his treatment of hypotheses in physics: there are two general means of uncovering rationally the real law of any given phenomenon—by induction, or the direct analysis of the sequence of the phenomenon, or by deduction, or establishing its exact and evident relation to some broader law, previously established.[50] The processes of logic are used in the objective synthesis in scientific observation, experimentation, and reasoning and in the subjective synthesis in education, inquiry, action, and religion. Physics is the science particularly suited to induction, and although deduction is also important, it is not dominant since the establishment of true principles is more embarrassing in physics than the development of just consequences. The simplicity of mathematical phenomena permits the establishment of solid principles by easy inductions. The true philosophic spirit is characterized more by induction than by deduction.[51] The method of induction is by observation or by experimentation; experimentation is fully suited only to inorganic and particularly physical researches, but it can also be extended to biology.[52]

Logic is not limited to intellectual formalism: it extends, on the one hand, from thought to emotion, to imagination, to language and, on the other hand, from reasoning about the hierarchy of phenomena to ordering the hierarchy of things. With respect to instrumentalities, different rational modes were applied in past ages to the elaboration of our abstract and general speculations—the force of sentiments, the efficacy of images, and the aptitude of natural and artificial signs. Indeed, since the end of the Middle Ages, logic has been restricted to the last of these three universal modes because it is proper to deduction even though it does not fit induction so well and popular use even less. The positivistic religion has established in the place of this vain separation of the logic of women or of the proletariat, the logic of the poets, and the logic of the philosophers and of the learned and indissoluble combination of all the regular means at the disposal of our nature to discover exterior laws.[53] With respect to matters, a twofold concatenation of the seven fundamental sciences, one objective and the other subjective, is established on scientific and logical grounds. In the objective hierarchy morals is subordinated to sociology (since the systematic study of man depends, logically and scientifically, on the study of Humanity), sociology to biology, biology to chemistry, chemistry to physics, physics to astromony, and astronomy to mathematics (since the geometric and mechanical phenomena of the heavens depend on the universal laws of number, extension, and motion). Logically the teaching of these subjects follows the order of their scientific relation. Deductive ability is acquired best by the simplest studies, while inductive attributes are acquired step by step with the study of more complicated phenomena—astronomical observation, physico-chemical experimentation, biological comparison, and sociological filiation. Once induction has completed deduction, the final science elaborates their normal combination directly, in constructing the subjective method, proper essentially to morals. The positivistic hierarchy may be used to represent the subordination of beings or existences one to another as well as the subordination of phenomena and speculations. In its concrete aspect the whole hierarchy constitutes a succession of states in which dignity increases with complexity. In this universal order, which is decomposed necessarily into seven categories superposed serially in such a way that each modifies the preceding and dominates the succeeding, man is represented as the normal resumé and the spontaneous regulator of the social, vital, and material milieu under which he develops.[54]

The controversies of Mill in 1865, concerning science and scientific method, provide the basis and materials for inquiry into the relations of science and philosophy before and after. Mill's statements expounding his position and the positions he opposed are univocal; they can be compared with the statements of his opponents. Semantic interpretation has two dimensions: in the semantics of philosophic controversy, the meanings given to terms in opposed statements are judged according to a single set of meanings, usually determined by the philosophy of the controversialist; in the semantics of philosophic history, the meanings given to terms in opposed statements are used to develop distinct sets of meanings, usually determining opposed philosophies. Common questions concerning science and philosophy take the place of Mill's univocal positions when the words he uses to express them are made to bear his opponents' meanings as well as his own. Philosophic inquiry, like philosophic semantics, has two dimensions: the end of inquiry is the solution of a specific problem stated univocally and consistently; the beginning of inquiry is the consideration of an ambiguous problem interpreted in alternative, mutually inconsistent hypotheses. The schematism remains constant in the semantic translation from one philosophy to another and in the transition from inquiry concerning a common ambiguous problem to inquiry concerning a specific interpretation of the problem. The structure that has been employed here in translating the nineteenth-century statement of controversy into a statement of inquiry concerning the relation of science and philosophy has been based on a matrix of philosophic speculations on the nature of science and the method of science. The rows of the matrix differentiated theories based on the structure of method or experience from theories based on the structure of science or existence; the columns differentiated the role of phenomena and the role of thought in science and scientific investigation. The history of the experiential theories after Mill and Whewell continued to be sharply distinct from the history of the ontic theories after Hamilton and Comte.

The difference between Mill and Whewell, which was expressed in their controversy concerning discovery and proof, was a difference concerning the place of facts and ideas in induction; induction was the basic process in both theories of scientific method. The difference between their classifications of the sciences was a difference between the use of induction to relate phenomena by causes on the one hand and to colligate facts by ideas on the other; physics and chemistry marked for both of them the point of crucial distinction of the sciences by the use of induction. Two other groups of sciences marked other points of critical variation in the history of experiential theories: mathematical and mechanical theories of the structure of phenomena and of the formation of proof, and biological and psychological theories of the genesis of man and of the structure of thought. Alexander Bain's *Logic: Deductive and Inductive* (1870) begins by examining the psychological data of logic, rounds out the treatment of deduction by setting forth "recent additions to the syllogism" by Hamilton, De Morgan and Boole,[55] and differentiates a logic for each of the sciences in classifying the sciences: a logic of mathematics, of physics, of chemistry, of biology, of psychology, of the sciences of classification, of practice, of politics, and of medicine.[56] Bain gives a brief history of classifications of the sciences which closes with a long critical account of Herbert Spencer's classification.[57] Spencer had written an essay, "The Genesis of Science," in 1854 to show that science cannot be rationally arranged in serial order as Comte had tried to do. In 1864, in "The Classification of the Sciences," he elaborated his distinction of science into science which treats the forms in which phenomena are known to us, or abstract science (logic and mathematics) and science which treats of the phenomena themselves in their elements, or abstract-concrete science (mechanics, physics, chemistry, etc.) and science which treats them in their totalities, or concrete science (astronomy, geology, biology, psychology, sociology). The third edition of that work in 1871 contained a reply to Bain's criticisms.[58] The tendency to seek the bases of induction in psychological laws of thought and sociological structures of value and to seek the structure of deduction in postulates of mathematics and the rules of art is illustrated clearly in the balance between John Venn's *The Principles of Empirical or Inductive Logic* (1889) and his *Symbolic Logic* (1881, revised edition, 1894), for in the first he starts with "the physical foundations of inference, or the world as the logician regards it" and "the subjective foundations of induction, or the principal postulates demanded on the mental side," and in the second he treats mathematical symbols and "the universe of discourse."

The difference between Comte and Hamilton, which was continued in the development of positivism and neocriticism, was a difference between conceptions of science as stating laws of phenomena and as applying laws of thought. The positivistic method and the idealistic logic were both sciences yielding opposed syntheses of abstract objective sciences and of concrete subjective actions. The difference between their classifications of the sciences was a difference in the relative importance given to thought and action in the development and synthesis of the sciences; the uses of the synthetic method were found in examination of the natural sciences and mathematics. During the second half of the nineteenth century the history of ontic theories was a history of the differentiation and comparison of *Naturwissenschaften* and *Geisteswissenschaften,* of natural science and history (Windelband, 1894) and natural science and cultural science (H. Rickert, 1899). Kant's critiques of pure and practical reason opened the way for Dilthey's critique of historical reason, Avenarius' critique of pure experience (*Erfahren*), and Husserl's critique of logical reason. The full title of Heinrich Rickert's *The Limits of Concept Formation in the Natural Sciences* (*Grenzen der Naturwissenschaftliche Begriffsbildung*) made it *A Logical Introduction to Historical Science* (1896–1902, fourth edition, 1921), and when his *Kulturwissenschaft und Naturwissenschaft* (1899, seventh edition, 1926) was published in English in 1962, it was given the title *Science and History: A Critique of Positivist Epistemology*. The work of Rickert had a major impact on the formation of Max Weber's theories. The universalities of the laws of natural science were set in relation to the particularities of history and of culture and the arts.

The experiential and the ontic theories evolved in the framework by which they are differentiated from each other, but they were so modified by their controversial oppositions and mutual influences that both encountered paradoxes which prepared for the revolutions in philosophy of the early twentieth century. In the experiential tradition, the examination of science and the differentiation of sciences by means of the inductive and deductive methods had moved to the examination of methods adapted to empirical data and formed to symbolic statement. The revolution was to pragmatic or operational logics of action on the one hand and to symbolic and mathematical logics of language on the other. In the ontic tradition, the classification of the sciences according to arrangements of underlying matters or according to specifications of supervening structures and the ordering of objective analyses by subjective syntheses had moved to the analysis of phenomenal experience and circumstantial existence. The revolution was to the analysis of existence and language on the one hand and to formal and transcendental logics of action on the other.

Empirical logics encountered paradoxes concerning the relation of true statements to what is the case, and symbolic logics encountered paradoxes concerning the relation of statements about things to statements about statements. The experiential tradition continued, despite paradoxes, to find the task of philosophy in relation to science to consist in the examination of the methods of science, and not in contributing to the establishment of scientific truths. Even though experimental logic was the method of discovering solutions to problems and mathematical logic could set up postulates from which all formalized sciences might be deduced, the methods of logic were themselves independent of metaphysical or epistemological commitment. The revolt was against idealism and to realism or naturalism. Phenomenological analyses encountered paradoxes concerning the relation of psychological and transcendental phenomenology, and formal logics encountered paradoxes concerning the relation of existence to being. The ontic tradition continued, with the aid of paradoxes, to find the task of philosophy in relation to science to be the constitution of philosophy as a rigorous science capable of treating objective problems, comparable to, but distinct from, those of the particular sciences, and not in speculation concerning abstract concepts and methods. Even though phenomenological analysis clarified experienced concepts and transcendental logic laid the foundations of formal logic, they neither established nor used chains of method but returned at each step to the immediately experienced. The revolt was against psychologism to phenomenology or existentialism.

The revolts of pragmatism and realism, phenomenology and conventionalism fifty or more years ago were radical revolutions. The marks of their break with the past are found in the terms and interests they shared—experience, existence, language, action, history, the concrete, facts, circumstances. Their revolutionary character appeared in their common tendency to discover errors in all previous statements of philosophic positions. Their conservative and philosophic character appeared in their common tendency to refute (or ignore) each other and to rectify, in a manner which Heidegger practiced to perfection, erroneous statements of past philosophic errors.

If the history of the revolts is treated as a history of philosophic positions, there are as many relations between philosophy and science as there are philosophies, and many of them are in complex controversial opposition on many points. If the revolts are examined as parts of a history of philosophic problems, the sharp division between philosophies before and after can be seen in terms which apply alike to science and philosophy and, in the philosophic evolution, to the experiential and the ontic traditions. In all alike, the paradoxes of late nineteenth-century controversy and discussion, inquiry and proof, had led to a change in the terms which formed the matrix of semantics and of inquiry. The nineteenth-century matrix had related the differentiation of method and science to the differentiation of phenomena and thought. The inductive method became dominant in the experiential tradition, and the synthetic method in the ontic tradition; the function of method in both traditions was to relate phenomena to principles. The paradoxes encountered in that enterprise suggested the reorientation of methodological inquiry to the facts of existence and experience rather than to the principles of systems. The matrix of semantics and of inquiry shifted from method and scientific system to interpretation and scientific statement, and from phenomena and thought to action and language. The logic of proof superseded inductive logic in the dominance induction had had in the experiential tradition, and the analytic method became the dominant method in the ontic tradition to replace the synthetic method of nineteenth-century philosophies. The revolution was from centering philosophic discussion of science on scientific method and order to centering it on scientific interpretation and explanation. Thought and what is are conclusions established from experience and fact rather than principles determinant of them. Within the new matrix the opposed traditions have shared the common characteristic, during the fifty years since the revolts, of discovering that some of the meaningless problems and statements of earlier philosophic discussion can be translated from the framework of methods to the new framework of interpretations and become meaningful or even unavoidable in the process.

The relation between philosophy and science, when

viewed as a common problem of philosophic inquiry rather than as particular consequences of philosophic positions, opens up new forms of philosophic semantics, philosophic history, and philosophic inquiry. Since philosophy and science treat common problems in inquiry, there is some indication that philosophic inquiry has begun again to learn from science to balance controversy concerning opposed positions (in which particular interpretations provide refutation of opposed theories) with inquiry concerning common problems (to which opposed theories provide alternative hypotheses of interpretation and solution). There are two dimensions of semantics: the semantics of a coherent philosophy used for the discovery of truth and the refutation of error, and the semantics of inquiry used to relate a variety of meanings to the interpretation of a common problem and to the comparison of divergent hypotheses. There are two related dimensions of history: the history of successive philosophic positions stated in the semantics of one coherent philosophy and the history of philosophic inquiry in the interpretation and transformation of common problems stated in the semantics which relates the divergent interpretations of an ambiguous problem, or statement, or fact. There are two dimensions of inquiry: inquiry which follows through the implications of a hypothesis and inquiry which examines divergent hypotheses even when they transform the original question to a different question related to different facts or to different interpretations of common facts. The revolutions of the twentieth century and the influence of science on philosophy and of philosophy on science give promise of a transformation of philosophy in all three respects: in the development of a semantics which relates philosophies by the interpretation of common problems; in the discovery of progress in the history of philosophy in the treatment of common problems; and in the advancement of inquiry in which different philosophic principles and theories are used as alternative hypotheses in the interpretation and modification of common concrete problems and situations.

NOTES

1. John Stuart Mill, *Auguste Comte and Positivism* (London, 1866) 54.
2. *Autobiography of John Stuart Mill* (New York, 1924), chap. 6, pp. 145–147.
3. Sir John Frederick William Herschel, *Preliminary Discourse on the Study of Natural Philosophy* (London, 1830), par. 303, p. 273.
4. Cf. R. McKeon, "Philosophy and the Development of Scientific Methods," *Journal of the History of Ideas* 27 (1966) 15–18.
5. Ernst Mach, *The Science of Mechanics: A Critical and Historical Account of Its Development,* T. J. McCormack, trans. (Chicago, 1907) 187. The italics are Mach's. In 1867 Kelvin and Tait expressed a like confidence that Newton's laws of motion stood in no need of alteration (Sir William Thomson and Peter Guthrie Tait, *Treatise on Natural Philosophy* [Cambridge, 1890], preface, p. vii): "In the second chapter we give Newton's Laws of Motion in his own words, and with some of his own comments—every attempt that has yet been made to supersede them having ended in utter failure. Perhaps nothing so simple, and at the same time so comprehensive, has ever been given as the foundation of a system in any of the sciences."
6. *Treatise on Electricity and Magnetism,* vol. 2, 3rd ed. (Oxford, 1892), part 4, chap. 5, art. 553, 554, and 567, pp. 199–200, 209–210.
7. *Regulae ad Directionem Ingenii,* Regula IV (*Oeuvres de Descartes,* vol. 10, C. Adam and P. Tannery, eds. [Paris, 1908] 377–378).
8. *Principia Philosophiae,* Pars II, 64 (*Oeuvres,* vol. 8, pp. 78–79).
9. *Meditationes,* Secundae Responsiones (*Oeuvres,* vol. 7, pp. 155–156). The French translation, which was seen by Descartes, translated *a priori* and *a posteriori* by the phrases in parentheses after those words (*Oeuvres,* vol. 9, pp. 121–122).
10. "Projet d'un Art d'Inventer," *Opusculus et Fragments Inédits,* L. Couturat, ed. (Paris, 1903) 181.

11. John Stuart Mill, *An Examination of Sir William Hamilton's Philosophy,* 3rd ed. (London, 1867), chap. 20, pp. 430 ff. and 459.
12. John Stuart Mill, *A System of Logic,* book III of Induction, 8th ed. (New York, 1893), chap. I, 1, p. 207.
13. *Autobiography,* chap. 7, pp. 157–158.
14. *A System of Logic,* book II, chap. IV, 1, p. 158; and 4, p. 162.
15. *Ibid.,* chap. IV, 5, p. 164.
16. *Ibid.,* book III, chap. I, 2, p. 208.
17. *Ibid.,* chap. III, 1, p. 208 and chap. XI, 1, p. 325.
18. *Ibid.,* book V, chap. IV, 3, p. 599.
19. *Ibid.,* book III, chap. VI, p. 167.
20. *Ibid.* 271.
21. *Ibid.,* book VI, chap. VI, 2, p. 608; chap. 9, 1, p. 619; chap. 10, pp. 630–631.
22. William Whewell, *On the Philosophy of Discovery* (London, 1860), preface, p. v.
23. William Whewell, *Novum Organon Renovatum,* 3rd ed. (London, 1858), preface, pp. iv–v.
24. *Philosophy of Discovery* v.
25. *Novum Organon Renovatum,* book II, chap. I, p. 27.
26. *Ibid.,* book III, chap. I, 2, p. 142.
27. *Ibid.,* chap. V, p. 186.
28. *Ibid.,* book II, chap. IX, 2, pp. 136–137.
29. *Ibid.,* book III, chap. X, pp. 247–256.
30. Sir William Hamilton, *Lectures on Logic,* lecture 4 (Boston, 1872) 41.
31. *Ibid.* 42–43.
32. *Ibid.,* lecture 17, pp. 225–228.
33. *Ibid.,* lecture 32, pp. 450–453.
34. *Ibid.,* lecture 27, pp. 377–378.
35. *Ibid.* 383.
36. *Ibid.* 379–382.
37. Auguste Comte, *Cours de Philosophie Positive,* vol. 1, 3rd ed. (Paris, 1869), Première Leçon, p. 8.
38. *Ibid.* 9–10.
39. *Ibid.* 16.
40. *Ibid.* 29.
41. *Ibid.,* Deuxième Leçon, p. 68.
42. *Ibid.* 71–75.
43. *Ibid.* 87.

44. Auguste Comte, *Système de Politique Positive,* vol. 1 (Paris, 1851), Discours Préliminaire, Première Partie, p. 8.
45. *Ibid.* 25–28.
46. *Ibid.,* Introduction Fondamentale à la fois Scientifique et Logique, pp. 443–444.
47. *Ibid.* 446.
48. Chapitre Troisième. Tableau Général de l'Existence Théorique, d'après la Conception Relative de l'Ordre Universel; ou Systématisation Finale du Dogme Positif (vol. 4 [Paris, 1854], 181–183).
49. *Ibid.* 187–189.
50. Auguste Comte, *Philosophie Positive,* vol. 2, Vingt-huitième Leçon, p. 298.
51. *Ibid.,* vol. 1, pp. 516–517.
52. *Ibid.* 651.
53. *Ibid.* 406.
54. *Ibid.,* vol. 4, pp. 182–185.
55. *Logic: Deductive and Inductive,* rev. ed. (New York, 1883) 178–207.
56. *Ibid.,* book V, "Logic of the Sciences," pp. 429–598.
57. *Ibid.,* appendix A, "Classification of the Sciences," pp. 627–639.
58. "The Classification of the Sciences," "Postscript, Replying to Criticisms," *Recent Discussions in Science, Philosophy, and Morals* (New York, 1886) 89–112.

3

THERMODYNAMICS AND RELIGION: A HISTORICAL APPRAISAL

Erwin Hiebert

I

The great bulk of literature on the interaction of science and religion in the Western world over the past century revolves largely about matters related to their warfare and reconciliation. In the majority of written works on this topic, since about 1850, the biases of the authors are transparent and strong in the defense of either science or religion. Thus, the losses to religion are chalked up on the credit side of science and vice versa. This attitude has prevailed in spite of the fact that many of the most intensely controversial issues seem to disappear into oblivion without any concessions of permanent damage in either camp.

In earlier times, by contrast, it was very commonly believed that an intimate knowledge of natural philosophy could only show the way to a greater reverence of the Deity who had called the cosmos into existence and sustained it. With such a concept of the presence and action of God in nature neither warfare nor reconciliation was a matter of vital concern for natural philosophers or theologians.

Within the mechanistic tradition of early modern science it was considered quite unacceptable to demand that God operate through fitful interventions from a supernatural sphere above and beyond the natural world. The alternative was to emphasize, as the deists did, that the study of nature revealed the wisdom of God in his original design. Until the middle of the nineteenth century nature was regarded as the most positive foundation for belief in the Deity. It was, in fact, a more acceptable foundation than the theology which had been the cause of persecution, war, and hostility since the Reformation.

In more recent times a favorite approach to the problem, viewed from the side of science, was to concentrate on those historical advances in the natural sciences which contributed most to the erosion of traditional religious institutions, practices, concepts, and beliefs. Frequently the intent was to demonstrate the practical absurdity and logical meaninglessness of specific elements of religious faith. This approach was founded on the assumption that the conflicts between science and religion were largely resolved by the advance of science and the retreat of religion.

A familiar argument proceeded somewhat as follows: Various aspects of nature, including vast domains of animate matter, once were thought to belong exclusively to the mysterious and sacred realm of truth which science would never invade. With the maturation of science modern man acquired factual information and scientific theories by means of which those "mysterious realms of truth" came to be annexed to science in various concrete developments within astronomy, evolution, genetics, meteorology, biochemistry, medicine, pharmacology, and the study of animal behavior.

It was claimed that as long as man's custom to worship God in a prescientific way prevailed, it was unlikely that man would examine the logic and meaning of his worship.

John Morley said it thus: "Where it is a duty to worship the sun it is pretty sure to be a crime to examine the laws of heat." The advancement of science changed these customs. With its unreasonably effective and precise mathematical language and its concepts about atoms, energy, genes, and cells, science appropriated to itself those realms of truth which once were taken to be the sole prerogative of religion. No wonder men were left hugging obsolete and empty religious concepts. No wonder the institutionalized rituals appeared sterile.

Religion according to this view was seen to augur a dim future except for those who were too witless to recognize the evidence against it and for those who were so persistent and ingenious in their mental gymnastics as to succeed in barricading themselves against scientific criticism. Religious systems based on naive stop-gap excuses for God in nature, the scientists argued, had provided their own built-in doom as science was seen to "take over."

Arguments of this type naturally led to the serious questioning of the values of the traditional religious beliefs and institutions. More detrimental to religion in the long run was the attitude of viewing traditional religion not as something to be combated or maligned, but just something fairly irrelevant to the life of man in society—something to be by-passed or quietly laid to rest while moving on to other nonsupernatural, less other-worldly, more secular tasks and objectives tied to more positive phenomenological evidence of what is and what might be.

Science itself was taken at times to represent a new high mode of religious activity—something akin to Spinoza's intellectual perception of God carried over into the life of the experiemential and theoretical scientist. Thus science was seen here to represent the search for God's truth, irrespective of the "beliefs" of truth's spokesman. At this level of discussion science and religion were no longer in competition with each other as a world view. "Winning out" and "conflict" had lost their meaning in the complete overlap of unitary objectives.

The alternative posture was to *accommodate* one's thought, to accept the mutually exclusive compartmentalization of science and religion by holding to two "complementary pictures" of man's view of the world: the scientific or material picture, and the religious or spiritual picture. Unfortunately the pictures were blurred, and the boundary conditions of the two pictures were ill defined. Thus no one could predict when and where the scientists would raise and answer questions which had been nurtured within the traditional boundaries of theology and philosophy. Always a possibility, to be sure, was the substitution of other ultimate systems of thought than those of history and tradition, that is, other religions.

At the opposite end of the problem, from the side of religion, we learn of efforts to reappraise and defend religion, theology, revelation, the church, and the "spiritual" needs of man in society by rejecting, denigrating, devaluating, or reinterpreting and adjusting scientific theories and explanations in ways designed to remove the existing conflicts. The literature devoted to this subject is enormous and covers the full range of arguments from shallow discourse, ignorance, intolerance, equivocation, and clever semantics to analyses and thoughtful encounter characterized by acuity, rationality, finesse, and charm.

The focus invariably fell on the removal, somehow, of irreconcilable dissimilarities and contradictions. For example, the scientific claims of mechanistic cosmology were taken to be exaggerated; literal biblical interpretations were seen to be too inflexible. Darwinian evolution was criticized for its endeavor to provide excuses for missing links. The biblical account of man's genesis was demythologized. The religious significance of biblical records, historical accounts, miracles, and textual contradictions were salvaged by reinterpretations which placed the emphasis upon nonliteral meanings of the text. In this way theology could secure itself against all the strictures of science by adopting a language which was not literal but symbolical and analogical. So too it could defend itself against the criticism of logic by claiming that its language was not logical but paradoxical. On the other hand, a knowledge of God which was merely symbolical, analogical, or paradoxical was considered by scientists to be no real knowledge at all.

II

Warfare or reconciliation, conflict or obsolescence, the broad historical approach to the interaction of science and religion is an engaging topic worthy of pursuit for the sake of reconstructing and evaluating man's social, cultural, and intellectual past. While there are a great many general works on this subject, I would suggest here today that we attempt a much more unconventional attack on the subject, viz., not to focus directly upon the conflicts and reconciliations between science and religion but rather upon the uses to which science has been put in answering questions of a fundamental religious cast.

Let me be much more specific and say that we want to concentrate exclusively on the influence which thermodynamics has exerted upon religious thought over the past century: we want to show how the first and second laws of thermodynamics were used, affirmed, rejected, manipulated, exploited, and criticized in order to further and also to censure religion.

As a historian of science, I find this a singularly exciting and provocative kind of study to engage in precisely because there are so many potential points of contact here between science and the development of modern thought. If the history of science is to be more than an internal analysis of the manner in which science came to be what it is today, or at any other time, then I suggest that the historian of science may have to venture out into such new territory. The new territory stretches out beyond the confines of the traditional study of the major documents which contributed to the experimental and theoretical progress of science. Indeed, if history is a record of man's ideas and behavior in place and time, then it may be quite as legitimate to study, for example, the interaction between thermodynamic thought and religion as it is to study the internal history of thermodynamics *per se;* and the latter is what I spend considerable time doing. However, the documents are quite different for the two kinds of historical study.

In our examination of the use of thermodynamic principles in religious (and perhaps pseudo-religious) thought, I recognize that we shall not always encounter what may with propriety be called first-class literature, either scientific or religious. Our criterion for considering a given author's treatment of a subject necessarily must be its relevance to history, and not its intrinsic merit as a work of enduring scientific or religious import. I am not simply defending the study of *any* crackpot thermodynamicians, philosophers, or theologians. The supply of such literature is immense. I am, however, suggesting the pertinence to this historical study of literary works which are crackpot or not, provided they exerted some noteworthy effect upon society. The works, in other words, must mirror something of the feelings and the attitudes of the age, irrespective of whether they would have stood the test of excellence in the eyes of the elite in science or religion, then as now.

The essential substance of this phase of the history of thermodynamics or religion, I will admit freely, is therefore not what happened but what was thought or said about it. I offer these comments by way of a kindly reminder of good intentions, just in case we might seem (as we move along) to be stressing the exotic, the nonsensical, or the irreverent. In analyzing history we need not always be profound, especially when the causes are so superficial.

III

I shall not discuss the history and meaning of the first two laws of thermodynamics except to identify them briefly within the context of our problem.

The first law of thermodynamics, the principle of conservation of energy, was enunciated in the 1840's by Mayer and Helmholtz in Germany, Joule in England, and Colding in Denmark. There were a number of other persons who entertained energy conservation ideas before and after the 1840's, but in one way or another they provided less satisfactory formulations of the principle than the persons already mentioned. Suffice it to say that conservation of energy was an independent multiplet discovery which burst forth among various European scientific investigators who were very much more closely tied to civil and military engineering, medicine, physiology, and brewing than to anything going on at the academic centers in the physical sciences.

In Julius Robert Mayer the law takes the form of an assertion that various forms of energy are qualitatively transformable, quantitatively indestructible, imponderable objects.[1] In the work of Mayer, Joule, Colding, and others we have, as well, theoretical calculations and experimental devices expressly designed to give the numerical equivalence of various forms of energy. Except for the rider to the first law, which Einstein attached in 1905 for mass-energy conversions, there are no clear-cut natural processes known to violate the principle of conservation of energy. In other words, it is impossible, by mechanical, chemical, thermal, or any other devices, to create any kind of perpetual motion machine which creates energy from nothing.

The second law of thermodynamics (an extension of Sadi Carnot's principle of 1824 and Clapeyron's algebraic and graphical representation of the same in 1834) was formulated in various ways by Clausius, Kelvin, Boltzmann, and others after 1850. This law extends beyond the principle of energy conservation by dealing with the direction in which a process can take place in nature, that is, energy conservation does not suffice for a unique determination of natural processes. For example, the second law stipulates how the exchange of heat by conduction can take place between two bodies of different temperature. The principle therefore has something to tell us about the quantity of heat which can be converted into mechanical energy in an ideal heat engine operating at a given temperature differential between the combustion chamber and the exhaust.

This means that we can talk meaningfully about available and unavailable energy. In fact, the capacity factor for unavailable energy is an extremely versatile thermodynamic function called entropy. It turns out that for all systems whose boundary conditions are defined in such a way that no energy transfers take place across the boundaries, the entropy increases for all spontaneous processes within the system. Entropy increase corresponds to a decrease in the available energy. The net result is that for all natural processes some of the energy ends up being unavailable. An equivalent statement would be to say that systems in nature move spontaneously from order to disorder, from lesser to greater randomness, or towards a state of maximum probability.

In 1865 Clausius put the first and second laws of thermodynamics into the following simple verbal form: "The energy of the universe is constant. The entropy of the universe tends towards a maximum."[2] These two statements along with various logical and illogical extensions of the energy and entropy principle constitute the background to all that I shall have to say here about the interaction of science and religion.

IV

The acceptance of the laws of thermodynamics, slow at first, especially among physicists and chemists, was followed by a period of rapid exploitation which revolutionized and unified the study of chemistry, heat theory, heat engines, radiation, electricity, and magnetism. Less respectable, but nevertheless real, were the far-reaching deductions, discussions, and speculations regarding the significance of thermodynamic concepts in cosmological works dealing with the source of the sun's energy, the origin of the solar system, and views on the expanding and contracting universe. Certainly, the energy concept also played an important role in discussions on the luminiferous, gravitational, and electrical ethers which were devised to deal with problems in electromagnetism, optics, radiation theory, and atomistics.

On the strength of the great wealth of far-reaching inferences which had been derived from a small number of axioms, thermodynamics by the end of the nineteenth century had taken its place alongside mechanics and electromagnetics as one of the main theoretical pillars of classical physics. Apart from concerns about the rate of chemical reactions, i.e., chemical kinetics, and the preoccupation with problems which arise in dealing with behavior of molecules and molecular aggregates, thermodynamics and thermochemistry, more than any other aspects of nineteenth-century science, contributed to the establishment of physical chemistry as a separate discipline.

At the turn of the century thermodynamics likewise aided substantially in the opening up of new fields of experimental and theoretical investigation, such as cryogenics, solid state physics, statistical mechanics, quantum theory, and the study of nonequilibrium, or irreversible processes. The fact that thermodynamics is no longer a logically necessary tool in dealing, for example, with quantum mechanics does not alter the fact that the early quantum theory was sheltered within the environment of thermodynamics.

It was virtually inevitable that some of the thermodynamic excitement of the nineteenth century should carry over into speculations concerning the problems of physiology, biological vitalism, and life in general. In truth, thermodynamics not only put its stamp on scientific thought but influenced, as well, social and political thought, psychology, literature, history, philosophy, and religion. It is the spilling over of thermodynamics into religion which we want to examine more particularly here.

Already in 1863 Sir William G. Armstrong in his presidential address to the British Association for the Advancement of Science remarked that the dynamical theory of heat and the new views on energy (or "force" as it was then still referred to loosely) probably constituted the most important discovery of the century.[3] By that time the principle of conservation of energy had been applied not only to motion, heat, light, electricity, magnetism, and chemical affinity but also to problems encountered in the world of organic life: animal and vegetable heat, digestion, respiration, muscular force, nervous agency, vital power, the development of organized animal and vegetable structure from dormant primordial germ cells, and the effects of plant and animal sensation and consciousness.[4]

One of the most widely read authors on this broad "philosophical" subject was the English lawyer and physicist William R. Grove whose famous *Correlation of Physical Forces,* first published in 1846, went through numerous English and American editions which kept the public informed on how the relentless march of the knowledge of various "forces" was solving the new problems of science.

Grove remarked that it was highly probable "that when discovered, and their modes of action fully traced out," these forces would "be found to be related *inter se,* and to these forces as these are to each other." This he believed "to be as far certain as certainty can be predicted of any future event." Grove concluded:

> Many existing phenomena, hitherto believed distinct, will be connected and explained: explanation is, indeed, only relation to something more familiar, not more known—i.e., known as to causative or creative agencies. In all phenomena

the more closely they are investigated the more are we convinced that, humanly speaking, neither matter nor force can be created or annihilated, and that an essential cause is unattainable. . . . Causation is the will, Creation the act, of God.[5]

William B. Carpenter, an eminent and unwearied English investigator in the sciences of zoology, botany, and mental physiology, wrote on the correlation of physical and vital forces with a gasconade which must surely have embarrassed the founders of the principle of conservation of energy. In a paper of 1850 Carpenter took up the problem of the mutual relation (metamorphosis and conversion) of the physical forces to the vital processes of plant and animal growth, development, reproduction, and evolution of complex heterogeneous structures from homogeneous germinal masses.[6] By applying the principle of correlation (conservation) of forces to vital phenomena he hoped, he said, to open up new lines of inquiry which would invest the physiological sciences with the same dynamic aspects as those under which the physical sciences were viewed by most enlightened philosophers.[7]

The particular *modus operandi* of "Cell-force" in the cell-formation of plants was compared to the "Engine-power," which term, Carpenter remarked, was used

> Knowing that the steam-engine possesses no power itself, but that it is simply that instrument most commonly employed, because the most convenient and advantageous yet devised, for the application of the expansive force of steam, generated by the application of heat, to the production of mechanical motion.[8]

Carpenter found the forces in the growth, development, and movement of animals to be essentially the same as "Cell-force" in plants except for an additional "Nervous agency" related to the conscious mind which communicates impressions derived from the external world and is related to the working of contractile tissues in obedience to mental impulses.

> For, just as electricity developed by chemical change may operate (by its correlation with chemical affinity) in producing other chemical changes elsewhere . . . so may nerve-force, which has its origin in cell-formation, excite or modify the process of cell-formation in other parts, and thus influence all the vital manifestations of the several tissues, whatever may be their own individual characters.

Thus he hoped to have established "the general proposition, that so close a mutual relation exists between all the vital forces, that they may be legitimately regarded as *modes* of one and the same force."[9]

As for the relations of the vital to the physical forces, Carpenter proceeded to demonstrate to his satisfaction that "Nervous Force" ("the *highest* of all the forms of vital force, both in its relations to mental action, and in its dominant power over organic processes of every kind") is perfectly correlated and mutually convertible with electricity, heat, light, magnetism, motion, and chemical affinity.

Nothwithstanding these physical comparisons, Carpenter retained, as Liebig had, the pre-existence of a living organism necessary for the development of an "organized structure of even the simplest kind."

> It is the *speciality* of the material substratum thus furnishing the medium or instrument of the metamorphosis, which . . . establishes and must ever maintain, a well-marked boundary-line between the Physical and the Vital forces. Starting with the abstract notion of Force, as emanating at once from the Divine Will, we might say that this force, operating through inorganic matter, manifests itself in electricity, magnetism, light, heat, chemical affinity, and mechanical motion; but that, when directed through organized structures, it effects the operations of growth, development, chemico-vital transformation, and the like; and is further metamorphosed, through

the instrumentality of the structures thus generated, into nervous agency and muscular power.[10]

Thus, for Carpenter, "all *force* which does not emanate from the will of created sentient beings, directly and immediately proceeds from the Will of the Omnipotent and Omnipresent Creator." What he calls "physical forces" are just "so many *modi operandi* of one and the same agency, the creative and sustaining will of the Deity."[11] Carpenter maintained reservations concerning the extension of the doctrine of evolution to man's intellectual and spiritual nature, even though he had moved by "grades of organic ascent" from an analysis of physical forces such as heat and light to nervous agency and muscular power.

V

We have indicated how the energy conservation principle was used in some fairly untamed, speculative, and theologically grounded arguments in the works of the physicist Grove and the physiologist Carpenter. In none of the early expositions of the energy principle does *philosophy* run riot quite so wildly as in the writings of Herbert Spencer. He was a representative of a widely disseminated naturalistic interpretation of phenomena which rested, in his case, upon a strange synthesis of ideas drawn from evolutionary theory, the principle of conservation of energy, and a humanitarian, religious, metaphysics. He was described by one of his friends as "radical all over."

In the *First Principles* of 1862 Spencer attempts a reconciliation of science and religion on the broad postulate of belief in the *Unknowable* as the cause of all phenomenal existence. The deepest, widest, and most certain of all facts, for Spencer, is that the *Power* which the universe manifests to man is utterly inscrutible. Thus "ultimate scientific ideas" such as force, space, and time are all representative of realities that cannot be comprehended. They pass all understanding.[12]

According to Spencer, among the ultimate scientific ideas "force" is the ultimate of ultimates and is rooted in primordial experiences. We do not know, he says, what force as an ultimate is, although the law of its manifestations can be inferred from experience—though never derived inductively. The law is the law of conservation of energy. "Persistence of force" is the expression which Huxley had suggested and which Spencer preferred.[13]

In his chapter on "the correlation and equivalence of forces" Spencer adds "mental forces" to the same generalization (the same law of metamorphosis) which he recognized as having been enunciated for physical forces on the basis of the experimental investigations of Mayer, Joule, Helmholtz and Grove.[14] Spencer says:

> Those modes of the Unknowable which we call motion, heat, light, chemical affinity, etc. are alike transformable into each other, and [also] into those modes of the Unknowable which we distinguish as sensation, emotion, thought: these, in their turns, being directly or indirectly transformable into the original shapes.
>
> Of course if the law of correlation and equivalence holds of the forces we class as vital and mental, it must hold also of those we class as social . . . [and] if we ask whence come these physical forces from which, through the intermediation of the vital forces, the social forces arise, the reply is of course as heretofore—the solar radiations.[15]

Spencer concludes by saying "the deepest truths we can reach are simply statements of the widest uniformities in our experience of the relations of Matter, Motion, and Force; the Matter, Motion, and Force are but symbols of the Unknown Reality."[16] Science, as in the Unknowable ultimate of ultimates, namely, *force,* coalesces with religion in the "consciousness of our Incomprehensible Omnipotent Power"—an Absolute Being. Religion and science are nevertheless both rooted in the common datum of all human thought, namely, in the law of the "persistence of force." But the establishment of the correlation and equivalence between the forces of the outer (matter) and the inner (spirit) world, for Spencer, are matters of the assimilation of either to the other, "according as we set out with one or the other term."

Related views on force are spelled out by Spencer in the section on "Ecclesiastical Institutions" in the third volume (*Principles of Sociology*) of his *Synthetic Philosophy* of 1885.[17] Here he is concerned with the history of religion and religious institutions and the elements of *power* that enter into all primitive religions. Acknowledging elements of power beyond consciousness, as evidenced in muscular power, ghost force, sun worship, the powers of medicine men and priests, Spencer concludes that all religions have a natural genesis which leads upwards from the most primitive anthropomorphisms to Gods, Inscrutable, Unknowable, and Omnipotent. The divinity which is synonymous with superiority eventually becomes nonanthropomorphic in the idea of force. This idea reaches its extreme form in the man of science who properly interprets the idea of the "persistence of force" in terms of all possible kinds of physical, biological, mental, and nervous phenomena.

Thus higher faculty and deeper insight, according to Spencer, raise the sentiments of the man of science to a vision, though dim and incomplete, of ultimate existence.

> And this feeling is not likely to be decreased but to be increased by that analysis of knowledge which, while forcing him to agnosticism, yet continually prompts him to imagine some solution of the Great Enigma which he knows cannot be solved. . . . But one truth must grow ever clearer—the truth that there is an Inscrutable Existence everywhere manifested, to which [man] . . . can neither find nor conceive either beginning or end. Amid the mysteries which become the more mysterious the more they are thought about, there will remain

the one absolute certainty, that he is ever in presence of an Infinite and Eternal Energy, from which all things proceed.[18]

The most enthusiastic apostle of Darwinism in Germany, a biologist of enormous literary output, the Jena Professor of Zoology Ernst Haeckel, applied the principle of conservation of energy (in combination with biological evolution) to some of the oldest problems in philosophy and religion. His *Die Welträthsel* of 1899, which appeared in English in 1900 as *The Riddle of the Universe,* adopted an inflexible, materialistic, monistic position which laid down the intrinsic unity of organic and inorganic nature and propounded the evolution of the highest level of the human faculties from unicellular protozoa. While Haeckel rejected the immortality of the soul, freedom of the will, and the existence of a personal Deity, he did not, he claimed, reject religion *per se*. His "monistic religion" with its "monistic church" was, he said, a monism connecting religion and science but freed from the dead and dried up superstitions of the traditional church religions.

In Haeckel's *Der Monismus als Band zwischen Religion und Wissenschaft* of 1892 we read that the most important general consequence of the spiritual conquest of modern science belongs to the law of substance (Substanz-Gesetz) —and that this is to be designated as the first paragraph of the "monistic religion of reason" (i.e., die monistische Vernunftsreligion). In a remarkable essay[19] of 1895 Haeckel wrote concerning this law:

> This supreme basic law of the cosmos actually consists of two intimately related laws: of the "law of the conservation of matter" for which we are indebted to the great French chemist Lavoisier, and of the "law of the conservation of energy" whose founding is shared by two German intellectual heros—the South-German Robert Mayer and the North-German Hermann Helmholtz. As "matter and energy" are inseparably combined in every thing, so also these two basic "conservation laws" hang together in one law of substance.

For the religion of reason of the science of today this law of substance is just as much the immovable foundation stone as the dogma of the "infallibility of the Pope" is for the Catholic church of today—"the rudest slap in the face for reason."[20]

Professor Haeckel ridicules the behavior of the mourners of the recently deceased Helmholtz. They behaved no better than Darwin's mourners did thirteen years earlier in the pantheon of Westminster Abbey. Haeckel asks whether the highly respected gentlemen who heard the church bells ring in Berlin for Helmholtz realized that they were honoring "a free-thinker who ought in their eyes to have been a revolutionary, mangy, heretic of the first order." Did none of them know that the greatest contribution of Helmholtz, the "law of substance" was paragraph number one of the "monistic religion" and that it hangs together inseparably with the infamous "materialistic" theory of evolution of Darwin?

VI

In Edward L. Youmans, the self-taught American promoter of science eduction, we have an ardent apostle of evolution and a devotee of the Spencerian philosophy. Youmans' influence on American scientific thought, in his role as a popular lecturer and as editor of the *International Scientific Series*, was truly impressive. This still on-going *Series*, which published over fifty volumes during his lifetime, included works by Liebig, Helmholtz, Darwin, and Huxley. Spencer apparently never delivered the manuscript he had promised. Youmans, who moved in the circles of William H. Appleton, Horace Greely, and Walt Whitman, also secured the establishment of the *Popular Scientific Monthly*.

Youmans' compilation of essays on *The Correlation and Conservation of Forces* of 1865 was published with the intent of demonstrating that scientific inquiries were shifting from questions about matter to questions of force. Material ideas were giving place to dynamical ideas.

> It has [been] shown that a pure principle forms the immaterial foundation of the universe. From the baldest materiality we rise at last to a truth of the spiritual world, of so exalted an order that it has been said "to connect the mind of man with the Spirit of God."[21]

Youmans felt that the law of conservation of "force" had opened up

> a region which promises possessions richer than any hitherto granted to the intellect of man . . . [having] pressed its inquiries into the higher region of life, mind, society, history and education.

The initiation of the "momentous event of intellectual progress" through the "establishment of a new philosophy of forces," Youmans felt, would bring about a perfectly correlated, marvelous, dynamic unity of the whole living system. The list of interconvertible forces included were: mechanical, thermal, luminous, electric and chemical forces, organic nutrition, muscular power, sensation force, nerve power, mental force, will power, cerebral force, brain force, emotional and intellectual force. Youmans wrote:

> And thus impressions made from moment to moment on all our organs of sense are directly correlated with external physical force. This correlation, furthermore, is quantitative as well as qualitative. . . . Between the emotions and bodily actions the correlation and equivalence are also directly correlated with physical activities. As in the inorganic world we know nothing of forces except as exhibited by matter, so in the higher intellectual realm we know nothing of mind-force except through its material manifestations.[22]

Youmans extended the force notion to include society and social economies as well.

> The law of correlation being thus applicable to human energy as well as to the powers of nature, it must also apply

to society, where we constantly witness the conversion of forces on a comprehensive scale.

Since according to the dynamical point of view there is a strict analogy between

> the individual and the social economies—the same law of force governs the development of both. . . . The amount of energy . . . is limited, and when consumed for one purpose it cannot of course be had for another. . . . So with the social organism; its forces being limited, there is but a definite amount of power to be consumed in the various social activities.

But the force law has yet higher bearings: history, the conditions of humanity, the progress of civilization, and the order of society in terms of the definite quantity of morality, justice, and liberty available. Societal progress thus depends upon the balance of these constantly changing ratios among the fixed quantity of these forces present, just as the individual growth of an organism depends on the balance of physiological, intellectual, and passional forces present. "So with society; the measured action of its forces gives rise to a fixed amount of morality and liberty in each age, but that amount increases with social evolution."[23]

And what is the place of God in this exuberant paean to the conservation of energy? The future of science opens up the infinite, mysterious, worship-worthy Cause of it all.

> And if these high realities are but faint and fitful glimpses which science has obtained in the dim dawn of discovery, what must be the glories of the coming day? If indeed they are but "pebbles" gathered from the shores of the great ocean of truth, what are the mysteries still hidden in the bosom of the mighty unexplored? And how far transcending all stretch of thought that Unknown and Infinite Cause of all to which the human spirit turns evermore in solemn and mysterious worship![24]

We might suggest an appropriate inscription for the Youmansian house of worship which connects "The mind of man with the Spirit of God": Let no man ignorant of the principle of conservation of energy enter here.

George Frederick Barker, another American scientist who was sometime professor of chemistry, geology, physiological chemistry, toxicology, and physics at Wheaton College in Illinois, at Yale, and the University of Pennsylvania, as well as president of the AAAS and the American Chemical Society, etc., accepted, without reservation, the philosophy of the correlation of vital and physical force and then managed somehow to twist the whole business to the glory of God. Barker wrote:

> I would fain believe that we now see more clearly the beautiful harmonies of bounteous nature; that on her many-stringed instrument force answers to force, like the notes of a great symphony; disappearing now in potential energy, and anon reappearing as actual energy, in a multitude of forms. I would hope that this wonderful unity and mutual interaction of force in the dead forms of inorganic nature appears to you identical in the living forms of animal and vegetable life which makes of our earth an Eden. . . . But here the great question rolls upon us. Is it only this? . . . Is there really no immortal portion separable from this brain-tissue, though yet mysteriously united to it? In a word does this curiously fashioned body enclose a soul, God-given and to God-returning? Here science veils her face and bows in reverence before the Almighty. We have passed the boundaries by which physical science is enclosed. No crucible, no subtle magnetic needle can answer now our questions. No word but His who formed us, can break the awful silence. In the presence of such a revelation Science is dumb and faith comes in joyfully to accept that higher truth which can never be the object of physical demonstration.[25]

VII

An unsystematic, nontheistic, monistic, and humanitarian exploitation of the philosophy of energeticism as a way of life was worked out and expounded in scores of books, pamphlets, and sermons by the famous German physical chemist Wilhelm Ostwald. In 1913, as president of the German Confederation of Monists, Ostwald delivered a lecture in Vienna which merits our attention.[26] This lecture admirably demonstrates the antireligious views of Ostwald with special reference to thermodynamics. Similar views were expounded at about the same time by Paul Carus, editor of the *Open Court* and *The Monist* in Chicago, and by August Wroblewski in Cracow, and Svante Arrhenius in Stockholm.

Briefly, Ostwald characterizes his brand of monism in this essay as a doctrine

> which excludes all double-entry bookkeeping, which removes all barriers, hitherto regarded as insurmountable, between inner and outer life, between the life of the present and that of the future, between the existence of the body and that of the soul, and which comprehends all these things in a single unity, that extends everywhere and leaves nothing outside its scope.[27]

Throughout man's past, ever since the pre-Socratic and Greek doctrines of the basic stuff of the universe, we discern, says Ostwald, man's attempt to struggle toward a unitary, monistic representation of the world. Thus, the history of philosophy is strewn with rejected monistic systems, including one that Ostwald also rejects, namely, the monism of a single supreme God. Ostwald's argument is that a unified self-consistent world view is unattainable by means of any *a priori* monism. This includes the monism of energy—as a principle. The only monism which Ostwald will recognize is an *a posteriori* monism "which proceeds from the diversity of the world as from the data offered by experience." Thus an *a posteriori* monism as the future ideal of unity is science. It is a monism of scientific thought and scientific method.[28]

Ostwald's program for scientific monism rests upon the

completion of a process which he finds to be already well under way, viz., the abolishment of the barrier or dividing line that separates the provinces of science and religion. This extremely movable barrier, Ostwald indicates, shifts invariably in the direction of enlarging and expanding the province of science at the expense of the province of religion, which thus grows narrower and narrower until reduced to almost nothing is forced to surrender. Thus Gustav Fechner's experiments, in Ostwald's opinion, have reduced the theological doctrine of the soul to a matter of measuring and recording. "Vainly, therefore," he says, "will the modern theologians still at the present day explain: "However far science may reach out, there is a point which it cannot reach, namely religious experience." Ostwald's answer to this: examine the work of the "pathfinding psychologist" William James who was at Harvard University. The struggle of the priests against science thus becomes one which "always allows us to discern the incessant and irresistible advance of science into the departments hitherto occupied by religion."[29]

As a "specimen of the practical Monism" which man should endorse, Ostwald mentions "the irresistible flow toward the international organization of human affairs." And what has made this possible? Answer: The movement toward great unity in all of science. Since when? Since the introduction and utilization of energy conservation into science.

This is Ostwald's so-called doctrine of the "energetic imperative."

> There must evidently be an imperative and decisive moment in the existence of thought, which gives to the principle of unity . . . so great an impulse that the whole tendency of human intellectual evolution surrenders to its grasp.

This movement "out of multifariousness to unity, especially out of the many dualisms to the one Monism" is, for Ostwald, symbolized by the "postulate of economy of energy."[30] In this manner Ostwald moves from scientific monism to the principle of energy conservation as the grand unifying principle of science, and finally to an energeticism as directive for the practical affairs of man in society.

Thus: "We know that we all live by virtue of the free energy that streams from the sun to the earth like a broad and powerful flood." Our energetic imperative must avowedly be:

> Waste no energy; turn it all to account. . . . Here, therefore, we have a genuine and far reaching Monism

which reaches from

> the simplest technical trade, yes from the daily acts of our half-animal life, to the very highest sociological and ethical problems.

Ostwald wrote:

> The whole culture-evolution which has led us from the invention of the slingstone, the lever, [and] fire . . . to a modern giant steamer . . . signifies nothing further than an always finer and more multiform manifestation of the energetic imperative. The same may be said of our moral evolution.

And thus, for Ostwald, the whole prophetic quality of science turns on the great synthetic unity of man's expenditure of his available energy.

> Accordingly, at the present time, the principal task of scientific or Monistic thought and labor is manifestly to free the final science in the succession of sciences, sociology, from the hitherto existing influence of the priesthood, and to establish in place of the traditional ethics dependent on revelation a rational scientific ethics, based on facts. . . .[31]

> Proceeding along this pathway of thought, we have come to recognize that the highest values of Christianity, the kindness and love of the individual toward his fellowman, do not yet

represent the highest ethical ideal which mankind can attain. Monism leads far more to the perception that the individual is more and more a mere cell in the collective organism of humanity. Accordingly, the evolution of kindness and love, the evolution of the spirit of self-sacrifice and devotion to the great whole of humanity, becomes more and more a demand of the energetic imperative, therefore an immanent demand of our whole Monistically ordered life. Only through the fact that we have come to recognize kindness and love as a necessity for community life, for the social organization of mankind, has there also been gained by the individual the sole sure and immovable foundation.

That we practice mutual kindness and love is no longer the demand of a Godhead standing outside of ourselves, which has once for all transmitted it to us by an unverifiable revelation; but it is a demand of scientific intelligence. To it, of course, only those can belong who dedicate themselves to Monism unreservedly and without any remnants of dualistic thinking and feeling. With the increased broadening and deepening of this intelligence we see arrive the Monistic century, which will not remain the only one of its kind. But it will inaugurate a new epoch for humanity, just as two thousand years ago the preaching of the general love for humanity had inaugurated an epoch.[32]

VIII

During the nineteenth century, the meaning of the law of conservation of energy was examined within the context of theological issues dealing primarily with the mode of God's presence and action in the world. The law was commonly invoked to bolster the argument for the existence of a Deity who had ordered the world with perfect foresight, wisdom, and economy of action. Thus matter (the stuff of the world) and energy (the action and process within the nature of things) were accepted as conserved quantities. They were built into a system of nature operating without over-all matter and energy losses. The argument for a world of law, order, and timeless permanence thus served theology very well as long as the laws of science were not seen to be so thoroughly successful in their account of the nature of things as to make God a mere benevolent, absentee landlord.

The philosophy of mechanism, based on the reduction of phenomena to matter and motion, had provided a serious threat to theism ever since the end of the seventeenth century. Indeed, deism was the bargain which religion had been disposed to make with mechanism. If God was not directly involved in the operations of nature, it was nevertheless still feasible to fall back on the belief that he had made it all in the first place. Unfortunately for Christianity the argument from design, with God as divine mechanic, was something quite unlike the God of the Bible.

With the seeming boundless accomplishments of nineteenth-century thermodynamics in explaining the mysteries of inanimate and animate process in nature, the energy concept, like the philosophy of mechanism, became a threat for those who were unwilling to relegate the description and resolution of the workings of nature, down to the smallest details, to the scientist. Thus, thermodynamics, with its analytical jurisdiction over conversions and redistributions of energy without losses and without divine interference, was no less tyrannical than mechanism had been. Energeticists like Ostwald were positively a fulmination to all that traditional religion had held in trust for centuries. Apart from Ostwaldian, monistic, and atheistic energeticism, some interpretations of the energy concept were viewed as injurious to questions concerning the possibility of miracles, the nature of free will, the creation and immateriality of the human soul, and the problem of death and immortality. There is an immensely spirited literature on this subject around the turn of the century, and we want to examine some of it briefly.

A systematic, theological statement on the energy problem came from the Thomistic or Scholastic study group which Pope Leo XIII encouraged at the University of Louvain in the 1880's. The newly created special chair in Neo-Scholastic philosophy was first occupied by the Belgian Cardinal Désire Joseph Mercier. It is the works of Cardinal Mercier and his school which will concern us here.[33]

We may in broad outline illustrate the approach to the interaction of thermodynamics and theology in the work of these Neo-Scholastics by mentioning the criticisms which were leveled against certain historic formulations of the principle of conservation of energy. For example, Helmholtz in his essay of 1862 on the conservation of energy had said that "the quantity of all the forces which can be put into action in the whole of nature is unchangeable and can be neither increased nor decreased."[34] Now the questions which one may raise in connection with such a broad statement are: How far does the energy concept reach into vital life processes, the motor or sensory nerve cells of the brain, problems of volition, consciousness, the nature of the human soul, the relation of the soul to the body, questions of moral freedom, and divine interference in the form of miracles? Cardinal Mercier would hold, for example, that Helmholtz's statement is a hazardous leap from positive science to very speculative metaphysics; that the statement is beyond the possibility of experimental proof.

While the energy conservation principle may hold for many systems, the Neo-Thomists deny that it has been satisfactorily demonstrated for physiological and psycho-physical systems, for consciousness, the human mind, and the moral law. Spencer's doctrine is erroneous because it converts the "persistence of force" into an absolute. Ostwald's views are false for rejecting matter and by so defining "energy" that it takes the place of "substance" in all processes of change—matter thereby becoming the mere capacity for the action of energy.

How about the conservation of energy in relation to the

human soul? Can the soul or the mind initiate or modify nature by resultant changes in matter and energy in the world? The Neo-Thomist answer to this question must be *yes*. The thoughts, feelings, and volitions of men must have some influence upon physical events. The soul must be able to act on matter; otherwise there can be no freedom of the will.

The next question is: *How* may the soul act on matter? By means of particular forms of energy interconvertible with other forms of energy? The answer is *no;* otherwise the soul is reduced to a form of energy or a mechanism. The soul therefore cannot act on matter by means of any guiding or directing energy, not even by means of a microscopic hair-trigger pressure movement. The chasms between the two orders of soul and matter are intellectually impassable.

Only a synthetic answer is possible for the mechanism of the soul's influence on physical events. The answer requires scientific information and religious interpretation, and the clue is given in the philosophy of Aristotle and St. Thomas. Thus Cardinal Mercier reawakens the Scholastic distinction between matter and form, where matter and form nevertheless exist only in conjunction. The "informing principle of substance" and "substance" coalesce in the specific existences which partake of the nature of things.

Maher wrote:

> The function . . . of this active informing principle is of a unifying, conserving, restraining character, holding back, as it were, and sustaining the potential energies of the organism in their unstable condition. From this view of the relation of the soul to the material constituents of the body, it would follow that the transformation of the potential energies of the living organism is accompanied in vital processes not by anything akin to positive physical pressure, but to some sort of liberative act. It would in this case suffice simply to unloose, to "let go," to cease the act of restraining, and the unstable forms of energy released will thereby issue of them-

selves into other forms. In a sac of gas or liquid, for instance, the covering membrane determines the contents to a particular shape, and conserves them in a particular space. Somewhat analogously, in the Scholastic theory the soul, as "form," determines the qualitative character of the material with which it coalesces while it conserves the living being in its specific nature.[35]

The theory thus avoids the dualism of body and soul. Body and soul act as a co-principle, with neither outside the other. The directing influence of the soul is not dependent upon the action on matter of something outside of it. "The soul is *in* the body which it animates and in every part of it."[36]

The claim of the Neo-Thomist doctrine for restricting the energy conservation principle to statements which cannot go beyond experimental proof (i.e., which cannot therefore extend into the domain of man's soul without reducing the soul to matter or energy) rests its case upon the construction of a synthetic philosophy which opens up the possibility of the soul's action on matter. Purportedly this solution therefore also solves some problems which otherwise remain unanswerable, such as freedom of the will, *creation* of the human soul, and the *departure* of the immortal soul. Neither *creation* nor *departure* is merely matter-energy collocation. Thomist philosophy therefore plays an important synthetic function in the matter of providing the meaningful synthetic interpretation of the energy concept. On the other hand, the Thomists insist that in order to give the synthetic explanations meaning presupposes a complete knowledge of the details furnished by the individual sciences.

I shall mention only the fact that various philosophers outlawed the Thomistic interpretation of the energy concept for religious, antireligious, monistic, materialistic, energetistic, materio-energetistic, scientific, and pseudo-scientific reasons. One of the most thorough-going criticisms, for example, came from the philosopher Konstantin Gutberlet who was very much preoccupied with the theological proofs for the existence of God. In 1885 he wrote a work entitled *Das Gesetz von der Erhaltung der Kraft und seine Beziehung zur Metaphysik*. In this work he attempted to support the substantial form of the energy notion with modern physical arguments.

Sir Oliver Lodge who had spoken on behalf of Christian orthodoxy against Haeckel's law of substance (conservation of matter and energy, conjointly) also expressed a view which was at variance with the Thomists. Lodge felt that matter and ether were substantial things, the only fundamental realities known to physical science. Energy or force, like movement, were mere abstract expressions. And life was not energy but a director of energy and matter. Vital force was nonsensical. The action of life on matter, Lodge nevertheless believed, was analogous to a hair trigger releasing large quantities of energy.

> So, inasmuch as liberation of energy can be accompanied by work entirely incommensurate with the result, it would appear that ultimately it can be achieved by no "work" at all—through the mysterious intervention of the brain as a connective between the psychic and physical worlds, which otherwise would not be in touch.[37]

Such a view was categorically rejected by the Thomists as an example of soul acting on body by mechanistic or energetistic means.

IX

The theological issues which were raised by the second law of thermodynamics were of a different character. If the principle of the increase of entropy of any operating system is extended to the world or the universe as a system, then the conclusion is reached that the universe is moving toward a state of maximum probability, a configuration of maximum randomness, a condition of minimum availability of energy, etc. This irreversible process, this unidirectional movement toward disorganization (time's arrow), leads to a degradation of the energy sources and ultimately to that pessimistic state of affairs, that boredom or Nirvana, which has been called the "heat-death" of the universe. Thermodynamics accordingly predicts an end to everything as a function of time.

Such a theory, which injected into the minds of men the consummation of the world as a majestic dysteleology, was bound to open up the flood gates of comments from religion. Ends and beginnings, the end of creation and the origin of creation—these are matters which theologians could hardly side-step. But what comes to an end? And what constitutes a beginning or a creation? I couldn't begin to review the answers given to such questions.

As representative of criticisms of the theological uses of the entropy principle, I suggest the published controversial correspondence which took place in 1933–1934 between Arnold Lunn and J. B. S. Haldane.[38] Lunn was a Madras-born novelist who was converted to Catholicism during the course of his correspondence with the evolutionist Haldane, who was then professor of genetics at University College in London.

In his discussions on "proofs of God's existence" Haldane chooses as his target for criticism the first proof of St. Thomas in the *Summa Theologica,* viz., the argument as to the unmoved mover. St. Thomas' philosophy, he concludes, is based on antiquated science and faulty mathematics—whatever that may mean. Then how about the modern argument (based on considerations of the second law of thermodynamics) which purports to demonstrate that the world cannot have existed for all times and which then raises the question whether the second law gives a sound new support for the theory of St. Thomas' first cause or unmoved mover?

Haldane finds the second law invalid for various "physical" reasons which we will not reconstruct here except in miniature. The universe may be spatially infinite so that the argument for finite systems fails. Even if finite, the universe may fluctuate back and forth between increasing and diminishing entropies which correspond to the running down and *also* the building up of the available energy. Or, in a space-expanding universe some apparently irreversible events (such as the radiation of heat to cold bodies) would be reversed in direction in a space-contracting universe, and so forth.[39] All of these physically feasible alternatives, we note, exclude the heat-death universe, and since they also do not support *processes* or *beginnings* and *creations,* therefore, says Haldane, by implication the proof of God as first cause or unmoved mover falls.

Lunn's reaction to the argument for God from entropy considerations was in substantial agreement with Haldane's criticism, but for different reasons. Here is Lunn's response:

> I . . . should like to hear Sir James Jeans reply. . . . I am reluctant to hitch the wagon of faith to the shooting star of scientific fashion. For all we know, relativity and the quantum theory and entropy will one day join their predecessors in the limbo of discarded scientific fads.[40]

Thus Haldane had no use for supernatural religion, while Lunn had no use for anything except rationalistic religion, which he employed to defend miracles and the supernatural, as well as to combat naturalism, materialism, evolution, Marxism, and Protestantism. Both seemingly agreed on the irrelevance of entropy considerations in the proof of God's existence: Haldane because he could not, at heart, believe in God; Lunn because he did not want his belief in God to be tied to the perennially obsolete fashions of science.[41]

Compare this with Eddington's view. On the one hand, he repudiated "the idea of proving the distinctive beliefs of religion either from the data of physical science or by the methods of physical science."[42] The nature of Eddington's conviction from which religion arises, while not founded in a rejection of reasoning (of the kind Lunn championed), nevertheless rested upon the belief that the ultimate data for reasoning was given through a nonreasoning process. According to Eddington this process was founded upon a state of self-knowledge or awareness of what is given in consciousness. For Eddington that was at least as significant as what is given in sensation.

On the other hand, Eddington had no such low opinion of the entropy principle as Lunn did. In his analysis of the relation of entropy to chance coincidences and random elements Eddington wrote:

> Entropy continually increases. We can, by isolating parts of the world and postulating rather idealized conditions in our problems, arrest the increase, but we cannot turn it into a decrease. That would involve something much worse than a violation of an ordinary law of Nature, namely an improbable coincidence. The law that entropy always increases—the second law of thermodynamics—holds, I think, the supreme position among the laws of Nature. If someone points out to you that your pet theory of the universe is in disagreement with Maxwell's equations—then so much the worse for Maxwell's equations. If it is found to be contradicted by observation—well, these experimentalists do bungle things sometimes. But if your theory is found to be against the second law of thermodynamics I can give you no hope; there is nothing for it but to collapse in deepest humiliation. This exaltation of the second law is not unreasonable. . . ; the chance against a breach of the second law (i.e., against a decrease of the random element) can be stated in figures which are overwhelming.[43]

Still another approach to the second law problem, radically different from that of Haldane, Lunn, or Eddington, was that of the archreactionary Christian Neoplatonist, the Dean of St. Paul's in London. The Very Reverend William Ralph Inge was a Christian mystic, a prolific writer of considerable influence, for whom the myth of progress, and evolution, were anathema. He was just as convinced that the fatal error of Catholic theology had been to find a rationalistic foundation for faith, as Lunn was convinced that Martin Luther was the father of the modern revolt against reason.[44] In his *God and the Astronomers* Inge undertook to demonstrate that any philosopher or theologian who wished to write on cosmology ("the relation of God to the universe") necessarily had to be informed on modern astronomy and physics. The research in these fields, he argued, could no longer be brushed aside as irrelevant to metaphysics or theology.

"Science and philosophy," Inge wrote, "can not be kept in watertight compartments."[45]

> God is revealing Himself to our age mainly through the book of nature. . . . This knowledge is given to us for a purpose. . . . Science has been called . . . the *purgatory* of religion. The study of nature . . . purifies our ideas about God and reality. It makes us ashamed of our petty interpretations of the world—ashamed of thinking that the universe was made solely for our benefit.

Thus Inge suggests that "the man of science worships a greater God than the average church-goer."[46]

Essentially, Dean Inge warned the theologians not to ignore the scientists. What scientists are doing, he said, is directly relevant to theological doctrines. For example, the religious and philosophical significance of the second law of thermodynamics—that new "Götterdämmerung," as Spengler calls it—has something to say about the ultimate fate of the world.[47]

Inge argues that if the universe is running down like a clock, the clock must have been wound up at some specific time. If the second law predicts an end in time for the world then the world must have had a beginning in time. Thus: "Is science itself driving us back to the traditional Christian doctrine that God created the world out of nothing?"[48]

If the second law leads to perdition, Inge is saying, then man must accept it. After all, it is a very Christian idea. The belief that the species should inherit the earth is hardly the doctrine of the Bible. The Bible does not forecast unending temporal progress, nor an evolving God, but rather the blessed hope of everlasting life. I quote the "gloomy Dean":

> Modern philosophy [with Time itself an absolute value, and progress a cosmic principle] is, as I maintain, wrecked on the Second Law of Thermodynamics; it is no wonder that it finds the situation intolerable, and wriggles piteously to escape from its toils.[49]

Dean Inge also finds the second law to be at odds with the theory of evolution. For since the universe is moving toward thermal equilibrium—its heat-death—both biological evolution as well as the idea of progress are illusions. Truly the species to which we belong has here on earth "no continuing city," and we should therefore "ascend with heart and mind" to "our citizenship . . . in heaven."[50]

The theological significance of the "end of man's history" view of things raised a host of questions which occupied religious scholars for decades. Was it consistent with God's goodness to annihilate creation through the heat-death? What meaning could then be attached to any values—in the mind of God—when man's history on earth would be terminated? For whom and in what way would these values be preserved and manifested? Under the circumstances of no other existence in the world than dead brute matter, would real existence reside only in God? If so, how could God's will and purpose be exercised at all? On what, if anything, might God then exercise his creative powers as an appropriate vehicle of his will? If creation is one of the symbols of the Creator, then in what way might an eternity of pessimism and doom represent an appropriate image of creation? Does "God," indeed, have any meaning in a universe which is not an abode for conscious life? And why should God create in order to destroy?

The second law would seem to be fatal to any meaningful view of the ultimate relation between the world and intelligent beings. Except, of course, within a metaphysical framework such as that provided by the Dean of St. Paul's. For he could supply answers such as the following: God is not limited to forms of expression which fall within the framework of man's knowledge. God can exist, if he wills, without men; for the world is not so necessary to God as God is to the world. "As a pleasure garden this world may be a failure; but it was never meant to be a pleasure garden."[51]

X

Finally, we want to examine what the scientist-philosopher-historian Pierre Duhem had to say about the relation between the second law of thermodynamics and religion. The occasion for Duhem's remarks on this subject were given in 1905[52] in the form of a reply and complaint to an article by Abel Rey.[53] In that article, while discussing Duhem's opinions concerning physical theories, Rey had implied that Duhem's scientific philosophy demonstrated that he was a believer.

Duhem wrote:

> Of course, I believe with all my soul in the truths that God has revealed to us and that He has taught us through His Church. . . . In this sense, it is permissible to say that the physics I profess is the physics of a believer.

But Abel Rey had meant more than that, viz.,

> that the beliefs of the Christian [Duhem] had more or less conspicuously guided the criticism of the physicist [Duhem]; that they had inclined his reason to certain conclusions . . . [which might] appear suspect to minds concerned with scientific rigor but alien to the spiritualist philosophy or Catholic

dogma, in short, that one must be a believer, not to mention being a perspicacious one, in order to adopt altogether the principles as well as the consequences of the doctrine that I have tried to formulate concerning physical theories.[54]

This Duhem denies. He is a Christian positivist and does not mind saying so. "Physics," he writes, "proceeds by an autonomous method absolutely independent of any metaphysical opinion." Theories have no "abilities to penetrate beyond the teachings of experiment or any capacity to surmise realities hidden under data observable by the senses." Theories have no more

> power to draw the plan of any metaphysical system [than] metaphysical doctrines [have] the right to testify for or against any physical theory. If all these efforts have terminated only in a conception of physics in which religious faith is implicitly and almost clandestinely postulated, then I confess I have been strangely mistaken about the result to which my work was tending.

Before admitting such a mistake, Duhem is anxious to fix his gaze particularly on those parts of his physics in which

> the seal of the Christian faith was believed noticeable, and to recognize whether against . . . [his] intention, this seal is really impressed therein or else, on the contrary, whether an illusion, easy to dissipate, has not led to the taking of certain characteristics not belonging to the work as the mark of a believer.

In any case, "the believer and non-believer may both work in common accord for the progress of physical science."[55]

The physical system which Duhem supports is positivist in its origins and in its conclusions. Physical theories are not metaphysical explanations. Nor are they a set of general laws whose *truth* is established by experiment and induction. Physical theories are artificial constructions mathematically relating abstract notions which emerge from experiment. They are "a kind of synoptic painting or schematic sketch suited to summarize and classify the laws of observation." Thus, as to their origins, the "reflections on the meaning and scope of physical theories were induced by preoccupations in which metaphysics and religion had no part." On the other hand, "they terminated in conclusions which have nothing to do with metaphysical doctrines and nothing to do with religious dogmas." It is crucial, therefore, to define "a theoretical physics for whose progress positivists and metaphysicians, materialists and spiritualists, nonbelievers and Christians may work with common accord."[56]

In the second place, Duhem's physical system "eliminates the alleged objections of physical science to spiritualistic metaphysics and the Catholic faith"; that is, it "denies physical theory any metaphysical or apologetic import." Accordingly—and here we come to our problem—a principle of theoretical physics can neither contradict nor confirm a proposition formulated in religion simply because "there cannot be disagreement or agreement between a proposition touching on an objective reality and another proposition which has no objective import."[57]

Thermodynamics as a physical principle is no exception. And this is what Duhem has to say about the second law in this context:

> In the middle of the last century, Clausius, after profoundly transforming Carnot's principle, drew from it the following famous corollary: The entropy of the universe tends toward a maximum. From this theorem many a philosopher maintained the conclusion of the impossibility of a world in which physical and chemical changes would go on being produced forever; it pleased them to think that these changes had had a beginning and would have an end; creation in time, if not of matter, at least of its aptitude for change, and the establishment in a more or less remote future of a state of absolute rest and universal death were for these thinkers inevitable con-

sequences of the principles of thermodynamics.[58]

Duhem suggests that such a deduction from the premises to the conclusions is marred by several fallacies:

> First of all, it implicitly assumes the assimilation of the universe to a finite collection of bodies isolated in a space absolutely void of matter; and this assimilation exposes one to many doubts. Once this assimilation is admitted, it is true that the entropy of the universe has to increase endlessly, but it does not impose any lower or upper limit on this entropy; nothing then would stop this magnitude from varying from $-\infty$ to $+\infty$ while the time itself varied from $-\infty$ to $+\infty$; then the allegedly demonstrated impossibilities regarding an eternal life for the universe would vanish.[59]

More important for Duhem is the objection based on "the very essence of physical theory," for

> We shall show that it is absurd to question this theory for information concerning events which might have happened in an extremely remote past, and absurd to demand of it predictions of events a very long way off.

Why is it absurd? Because physical theories are mathematical propositions to represent the data of experiment; and "if two different theories represent the same facts with the same degree of approximation, [then] physical method considers them to have absolutely the same validity." So the scientist then is free to choose between the logically equivalent theories from considerations "of elegance, simplicity, and convenience, and grounds of suitability which are essentially subjective, contingent, and variable with time, with schools, and with persons."[60]

Thermodynamics too, in Duhem's system, employs arbitrary theories which are adequate to the job at hand. The predictions of such theories will merit a certain degree of confidence, but their logic gives no right to assert that the predictions of these theories, and not others, will be in conformity with reality. Thus with respect to any long-term predictions the principles of thermodynamics have no unambiguous truth status.

Duhem says:

> We possess a thermodynamics which represents very well a multitude of experimental laws, and it tells us that the entropy of an isolated system increases eternally. We could without difficulty construct a new thermodynamics which would represent as well as the old thermodynamics the experimental laws known until now, and whose predictions would go along in agreement with those of the old thermodynamics for ten thousand years; and yet, this new thermodynamics might tell us that the entropy of the universe after increasing for a period of 100 million years will decrease over a new period of 100 million years in order to increase again in an eternal cycle.
>
> By its very essence experimental science is incapable of predicting the end of the world as well as of asserting its perpetual activity. Only a gross misconception of its scope could have claimed for it the proof of a dogma affirmed by our faith.[61]

In conclusion: What does the scientist need to know about metaphysics in order to do science? I believe Duhem's answer must be *nothing*. And what does the metaphysician need to know about science in order to do metaphysics? Duhem's answer is: "The metaphysician should know physical theory in order not to make an illegitimate use of it in his speculations."[62]

XI

There are a great many additional problems on the interaction of thermodynamics and religion which have not been explored here at all. One of the most real of these—at least to judge from the quantity of literature on the subject—concerns the question of how thermodynamics and evolutionary theory, in the nineteenth century, were to be viewed by and correlated with religion at the same time.

The knowledge of the second law of thermodynamics was in the general scientific domain while some of the enthusiastic supporters of biological evolution were indulging in the most jubilant song of progress of the century. In fact the mood of sympathy for movements toward greater perfection, which was often read into evolution, lay in the atmosphere of most of the religiously enlightened persons during the Victorian reign, not to speak of its infectious spread into the literature of philosophy, aesthetics, and the political, social, and historical sciences. But how was the God of evolution, the proud master of a vital eternally progressing universe, to be squared with the God of the second law—the lonesome remnant of a dead universe? Perhaps the Buddhists with their doctrine of Nirvana could have embraced the law of degradation, but certainly not the energetic progress-happy Victorians.

Evolutionary theory was equated with a century of hope, the law of progress, the lay creed of men of science. Entropy considerations prophesied a pessimism of despair, a future of disillusionment. It is something of an enigma to see that Spencer could feel so empathic with the energetic concepts and simultaneously identify progress with the increasing complexity offered through evolution. I suspect he never comprehended the second law. Or perhaps he had no time to read the literature; he was either busy writing or being ill.

Energeticists such as Ostwald stressed those aspects of the second law of thermodynamics which dealt with the spontaneous irreversibility of time and what that signifies for man's progress. What Ostwald had in mind, naturally, was the element of irreversibility and progress in the development of culture, aesthetics, morals, and man's store of knowledge. This was the inheritance of the human race resulting from the accumulative and irreversible progress of man. This is an exploitation of time's arrow which does not dwell on the final dismal outcome of things. Ostwald played down the gloomier aspects of the second law by ignoring them completely.

References to thermodynamics and religion—some oblique, some not so oblique, and some quite blunt—are to be found in the writings of many scientists. Mayer, Joule, and Colding included theological comments along with their formulations of the first law. Tyndall, the Irish physicist, attributed to the doctrine of the conservation of energy a wider grasp and a more radical significance than to the theory of the origin of species. Energy conservation, he thought, "bound nature fast in faith" by bringing vital as well as physical phenomena under the dominion of causal connection. He wrote: "If our spiritual authorities could only devise a form in which the heart might express itself without putting the intellect to shame, they might utilize a power which they now waste, and make prayer, instead of a butt to the scorner, a potent supplement of noble outward life."[63]

Kelvin excluded living organisms in his statement of the second law. Arrhenius, the Swedish chemist, while admitting the universality of the second law, thought that rare exceptions might occur leading to the rebuilding of worlds, perhaps by some Maxwellian conscious demon. Meyerson concluded that the second law was a fatal obstacle to the mechanical explanation of the universe.

The philosophies of Nietzsche, Whitehead, and Russell are not inconsequent on these problems. Bergson called the second law the most metaphysical of all the principles of nature because of its implication of a beginning in time. Freud and Jung took a new look at religious experience with the help of wild ideas and brand new expressions: the energy laden world waiting to react on man, traumatic neuroses in relation to bound "psychic energy," the Ich-

Energie as narcistic libido or desexualized eros, cathectic energy, entropy, and the death wish. William James wrote:

> Though the ultimate state of the universe may be its vital and physical extinction, there is nothing in physics to interfere with the hypothesis that the *penultimate* state might be the millenium. The last expiring pulsation of the universe might be—I am so happy and perfect that I can stand it no longer.[64]

Henry James thought the universe had been so terribly narrowed down by thermodynamics that history and sociology had to "gasp for breath." The worship of the dynamo had replaced the worship of the Virgin of Chartres. Perhaps Chesterton was right when he said that the test of a good religion is whether or not you can make a joke about it. Or, it all depends upon the point of view.

XII

I have attempted one possible approach to the historical study of the interaction of science and religion, namely, with the emphasis upon some of the uses to which science was put in discussions of religious issues. In this case thermodynamics was chosen for analysis.

Obviously, this is not at all the same kind of mental exercise as the internal history of science, or of religion. Nor is it the study of the history of science or religion as a branch of intellectual thought; nor explication, nor hermeneutics. It has something in common with social and cultural history, perhaps *Kulturgeschichte*. The question is: Do we have anything to learn from this exciting, sometimes crazy, but widely read literature?

Certainly we recognize that persons once shared a marked degree of religious frenzy about the importance of thermodynamics for "ultimate" questions about man's life. If this now seems to us more rash than was warranted, more subjective than was proper, and more trusting in the incorruptibility of truth than seems decent, it was nevertheless quite real. I suggest that this was due, in part, to the fact that the evolution of thermodynamics was particularly susceptible to logical insecurity. I mean that thermodynamics was formulated before all the essential physical facts were in. There were too many exciting overtones to postpone the triumph. Thus the laws of thermodynamics, adorned the more by their great generality and invulnerability, seemed to have sprung full bloom from the head of Zeus.

Conservation of energy was virtually accepted as the mere extension of the impossibility of a *perpetuum mobile* to all processes in nature—or so it seemed in retrospect. The second law was not so obviously on the surface of things. Entropy was, as Emile Meyerson saw, a less plausible, a more irrational, concept. The first law was almost a triumph of *a priori* human thought. The second law lay concealed not only beneath steam power engineering and thermodynamic efficiency but beneath coincidence, chance, probability, equilibrium and irreversibility. Still, once formulated, the second law was very much more loaded with potential interpretations which were relevant to matters of concern in religion.

In any case, we have here the record of how science was put to use in the service of religion. And the record seems to demonstrate the general verdict: that a scientific theory, no matter how good, secure, and elegant, is never immune to being used in such a way as to transgress the limits of credulity to the point of sheer ridiculousness, at least in the eyes of subsequent generations.

Is it possible at all to manufacture a metaphysics by the methods of, or with the aid of natural science? How does one use science to pay God metaphysical compliments? The physical Weltanschauung of mechanism and energeticism were all once taken seriously; and now are more or less forgotten. We note, in fact, that all kinds of private metaphysics have grown like weeds in the garden of thermodynamics. It was inevitable that the second law, which threatened at one point to destroy the universal validity of mechanism in the world, was worthy of serious attention by philosophers, metaphysicians, and theologians. It didn't last. And so we might legitimately ask whether religion and science must always have relations with each other. Not very much, says Duhem. If the answer is yes, then we must ask: What kind of relations? And how *is* science built into the ceremony and life of religion?

Religion is vastly older than science and once held the field alone without great competition. Is it conceivable that scientific criticism, reinterpretation, sophistication, could put an end to religion? In the relative vacuum created by the crumbling of religious orthodoxy in the nineteenth century something was bound to take its place—i.e., Spengler, Ostwald, or Thomism. Science had left a great many persons holding empty beliefs. And, is it not noteworthy to recognize how persons who ridiculed the Christian faith were often oblivious about their own assumptions, difficulties, contradictions, speculations, and synthetic monstrosities?

I conclude with the observation that the substantive documents which served as the basis for this study were imbedded in words like *God, creation, world, origins, time, energy,* and *entropy.* These were used with so many different meanings in the literature that it should not surprise us completely that one and the same set of principles could be manipulated to demonstrate such diverse theses. It should remind us of Wittgenstein's dictum: "A philosophical problem has the form: 'I don't know my way about.' . . . For philosophical problems arise when language *goes on holiday.*"

NOTES

1. J. R. Mayer, "Bemerkungen über die Kräfte der unbelebten Natur," *Annalen der Chemie und Pharmacie* 42 (1842) 234.

2. Rudolph Clausius, "Ueber verschiedene für die Anwendung bequeme Formen der Hauptgleichungen der mechanischen Wärmetheorie," *Annalen der Physik und Chemie* 125 (1865) 400.

3. Sir William G. Armstrong, "Presidential Address," *British Association for the Advancement of Science*. Annual Report of 1863, London, 1864, li–lxiv.

4. See Edward L. Youmans, *The Correlation and Conservation of Forces* (New York, 1865), a series of expositions by Prof. Grove, Prof. Helmholtz, Dr. Mayer, Dr. Faraday, Prof. Liebig, and Dr. Carpenter; Thomas Laycock, *Mind and Brain: The Correlations of Consciousness and Organization,* 2 vols. (Edinburgh, 1860), with their applications to philosophy, zoology, physiology, mental pathology, and the practice of medicine; James Kay Shuttleworth, "On public education and the relation of moral and physical forces in civilization," *National Association for the Promotion of Social Science, Transactions* (London, 1861) 79–109; James Hinton, *Life in Nature* (London, 1862). In Germany and Holland the public response to the philosophical and theological significance of the energy conservation principle was stimulated by discussions about vital force (Lebenskraft), the matter-force doctrines of the materialists, and hostile reactions thereto. See, e.g., Justus Liebig, *Chemische Briefe* (Heidelberg, 1845); Carl Vogt, *Physiologische Briefe* für Gebildete aller Stände, 2nd ed. (Giessen, 1853); Jacob Moleschott, *Der Kreislauf des Lebens. Physiologische Antworten auf Liebig's Chemische Briefe* (Mainz, 1852); Ludwig Büchner, *Kraft und Stoff, Empirisch-naturphilosophische Studien* (Frankfurt, 1855); Friedrich Lange, *Geschichte des Materialismus und Kritik seiner Bedeutung der Gegenwart* (Leipzig, 1866); and numerous revisions of all of the above works.

5. William R. Grove, "The Correlation of Physical Forces," quoted from the first American edition as given in Youmans, *The Correlation and Conservation of Forces* 178, 199.

6. William B. Carpenter, "On the Mutual Relations of the Vital and Physical Forces," *Phil. Trans.* 140 (1850) 727–757.

7. Carpenter, *ibid.* 757.

8. *Ibid.* 737.
9. *Ibid.* 740–741.
10. *Ibid.* 752.
11. *Ibid.* 730.
12. Herbert Spencer, *First Principles of a New System of Philosophy,* New York 1866 edition of the first published version of 1862, pp. 44, 66.
13. Spencer, *ibid.* 235.
14. *Ibid.* 259.
15. *Ibid.* 280–282.
16. *Ibid.* 501.
17. Herbert Spencer, "Ecclesiastical Institutions," part VI of the *Principles of Sociology* (New York, 1866) 827–843.
18. Spencer, *ibid.* 843.
19. Ernst Haeckel, "Die Wissenschaft und der Umsturz," *Die Zukunft* 10 (1895) 197–206.
20. Haeckel, *ibid.* 199.
21. Youmans, *The Correlation* xii.
22. *Ibid.* xxx–xxxv.
23. *Ibid.* xxxvi–xxxvii.
24. *Ibid.* xlii–xliii.
25. George F. Barker, *The Correlation of Vital and Physical Forces* (New Haven, 1870) 26–27.
26. Wilhelm Ostwald, *Monism as the Goal of Civilization* (Hamburg, 1913). Published by the International Committee of Monism and distributed as a brochure free of charge. For a sarcastic analysis of Ostwarld's brand of energetistic monism see Oskar H. Schmitz, *Die Welt der Halbgebildeten,* 6th ed. (Munich, 1914); also Wilhelm von Schnehen. *Energetische Weltanschauung? Eine kritische Studie mit besonderer Rücksicht auf W. Ostwalds Naturphilosophie* (Leipzig, 1908).
27. Ostwald, *Monism as the Goal of Civilization* 5–6.
28. *Ibid.* 6–19.
29. *Ibid.* 20–23.
30. *Ibid.* 25–26.
31. *Ibid.* 30–35.
32. *Ibid.* 37.
33. Michael Maher, "Energy," *The Catholic Encyclopedia* 5 (1909) 422–428; D. Mercier, *La pensée et la loi de la conservation* (Louvain, 1900); M. P. De Munnynck, "La conservation de l'énergie et la liberté morale," *Revue Thomiste* 5 (1897)

153–179; D. Mercier, *A Manual of Modern Scholastic Philosophy* (London, 1917).

34. Hermann von Helmholtz, *Vorträge und Reden,* vol. 1 (Braunschweig, 1884) 152.

35. Maher, "Energy" 427–428.

36. *Ibid.* 428.

37. Oliver Lodge, *Man and the Universe* (London, 1909) 69. For a comprehensive discussion of the role of energy conservation in the interaction between the mental and the material aspects of things, see the notes by Lodge, Hobson, Minchin, McDougall, Worthington, Sharpe, Peddie, Preece, Muirhead, Culverwell, and Bowman in *Nature* 67 (1903) 595–597, 611–612; 68 (1903) 31–33, 53, 77–78, 126–127, 150–152.

38. Arnold Lunn and J. B. S. Haldane, *Science and the Supernatural* (New York, 1935), a correspondence.

39. Lunn and Haldane, *ibid.* 167–175.

40. *Ibid.* 261.

41. See, e.g., Arnold Lunn's "Introduction" to D. Dewar and H. S. Shelton, *Is Evolution Proved?* (London, 1947); Arnold Lunn, *The Revolt against Reason* (New York, 1951); J. B. S. Haldane, *The Outlook of Science* (London, 1935); J. B. S. Haldane, *Science and Well-Being* (London, 1935).

42. Arthur Eddington, *The Nature of the Physical World* (Cambridge, 1928) 333.

43. Eddington, *ibid.* 73–75.

44. W. R. Inge, *The Church in the World* (London, 1928); Arnold Lunn, *A Flight from Reason* (London, 1931), a Study of the Victorian Heresy.

45. William Ralph Inge, *God and the Astronomers* (London, 1934) v–vii, the Warburton Lectures of 1931–1933.

46. Inge, *ibid.* 15–16.

47. *Ibid.,* chap. 2, "The New Götterdämmerung" 19–70.

48. *Ibid.* 10.

49. *Ibid.* 25–28.

50. *Ibid.* 127.

51. *Ibid.* 33.

52. Pierre Duhem, "Physique de Croyant," *Annales de philosophie chrétienne* [4] 1 (October and November 1905) 44 ff. and 133 ff. Reproduced in Duhem's *Aim and Structure of Physical Theory* (Princeton, 1954) 273–311, appendix, "Physics of a Believer."

53. Abel Rey, "La philosophie scientifique de M. Duhem," *Revue de Métaphysique et de Morale* 12 (1904) 699–744.
54. Duhem, *Aim and Structure* 273–274.
55. *Ibid.* 274–275.
56. *Ibid.* 275–279.
57. *Ibid.* 282–287.
58. *Ibid.* 287–288.
59. *Ibid.* 288.
60. *Ibid.* 288.
61. *Ibid.* 290.
62. *Ibid.* 291.
63. John Tyndall, *Constitution of the Universe* (Melbourne, 1869) 32.
64. Inge, *God and the Astronomer* 30. Quoted from Urban, *The Intelligible World*.

4

SCIENCE A CENTURY AGO

Michael Crowe

When, nearly two thousand years ago, Titus Livy began his history of Rome, he did this in the hope of distracting himself from the "melancholy spectacle" of his age by reflecting on the heroes of the past. When, nearly three hundred years ago, Isaac Newton (presumably following Bernard of Chartres) commented, "If I have seen farther, it is by standing on the shoulders of giants," he was referring to scientists some of whom lived hundreds of years earlier. Scientists of the present day cannot share the sentiments of Livy, for they live in the golden age of science; and they have difficulty in sharing Newton's feelings, for they frequently are contemporaries of the giants on whose shoulders they stand. Nonetheless, on the occasion of the Centennial of Science of the University of Notre Dame, it is especially appropriate for us to turn to that decade which now lies one century distant and to acknowledge our debt to the giants of that age.

In what follows, an attempt is made to set forth the major scientific discoveries which came in the 1860's. Various limitations have necessitated that the focus be limited to discoveries made in mathematics, physics, chemistry, and biology. However, it has seemed wise to include some theories that were propounded in the 1860's but are no longer held. This may serve to remind us of the important truth that the scientists of the past not only saw less far but also saw differently.

MATHEMATICS IN THE 1860's

George Cantor once remarked, "The essence of mathematics lies in its freedom." If this is true, then the discovery of the essence of mathematics may in a very real sense be associated with the 1860's, for in that decade mathematicians recognized more fully than ever before the wonderful freedom of creation that mathematics allows. Two series of developments, one geometric and one algebraic, both of which culminated in the 1860's, led to this realization.

In 1829 the Russian Nicholaus Lobachevski published a system of non-Euclidean geometry, and four years later the Hungarian Johann Bolyai, working independently, published a similar system. However, the writings of Bolyai and Lobachevski attracted no attention. Carl Friedrich Gauss had discovered non-Euclidean geometry slightly earlier than Lobachevski and Bolyai, but Gauss had not published his results. In 1854 Bernard Riemann delivered a lecture in which he described a new system of non-Euclidean geometry, but Riemann's paper was not published at that time. Thus by the beginning of the 1860's non-Euclidean geometry, including n-dimensional geometry, had been discovered, but few mathematicians knew of these ideas. This situation rapidly changed, for in 1860, shortly after Gauss's death, his views were made known by the publication of some of his letters; and the two-year interval 1866–1867 saw the translation and publication in France of the treatises of Bolyai and Lobachevski, as well as the first publication of Riemann's lecture. Important publications by Helmholtz and Beltrami appeared before the end of the decade, and by 1870 most mathematicians had recognized the wonderful geometrical freedom inherent in their domain.

The dawn of algebraic freedom came at nearly the same time. The acceptance by mathematicians of complex or imaginary numbers is usually dated from Gauss's publication, in 1831, of the geometrical representation of complex numbers. Beginning in the 1840's, this extension of the idea of numbers was followed by the creation of higher complex numbers, most notably by the Irishman Hamilton and the German Grassmann. The earlier complex numbers, though certainly shocking, did not necessitate the abandonment of any of the traditional laws of algebraic operation. Such was not the case with the systems of Hamilton and Grassmann wherein $A \times B$ does not in general equal $B \times A$. But the ideas of Hamilton and Grassmann were not at first well received; indeed, many copies of Grassmann's book of 1844 went unsold and were used as waste paper.

By the 1860's the situation was changing, and the work of Benjamin Peirce may be cited to symbolize this change. In 1864, at the first meeting of the newly founded National Academy of Science, Peirce read the first of a series of papers, which in 1870 were brought together in his "Linear Associative Algebra" of 1870. Herein Peirce set forth no less than 162 separate mathematical systems.

With this newly found algebraic freedom came such branches of mathematics as vector analysis and matrices. Vector analysis in a sense dates from the 1860's when the earlier work of Hamilton and Grassmann was taken up by such men as Tait, Maxwell, Clifford, Schlegel, and Hankel. Matrices, though adumbrated by Hamilton and especially Grassmann, are generally associated with Cayley's paper of 1858.

Also in the 1860's, and partly as a result of the researches already mentioned, mathematicians came to realize the importance of studying not only mathematical entities but also mathematical structures. The major tool for this task was group theory, which was begun by Evariste Galois who died in 1832 at age twenty. Galois' ideas began to receive attention only after 1846 when Liouville published Galois' papers. The culmination point in this line of development is usually taken to be the publication in 1870 of Camille Jordan's book on substitution groups. Two young mathematicians who attended Jordan's lectures in 1870 were destined to carry the development of group theory much further. They were Felix Klein, whose now classic Erlanger program appeared in 1872, and the Norwegian Sophus Lie, whose contributions to group theory are likewise classic.

That analysis was not negelcted in the 1860's may be established by the mention of such mathematicians as Riemann, Hermite, Kronecker, and above all Weierstrass. Two developments, far-reaching in their implications, are attributed to Weierstrass's lectures in the 1860's. Weierstrass at that time showed the existence of continuous functions which have no derivatives at any point, and he thus inaugurated a crisis in mathematical intuition. Four years later Weierstrass told his students of his method for developing the irrational numbers on the basis of the integers. This began the "arithmetization of analysis," and Weierstrass' work was soon followed by that of Méray (1869), Dedekind (1872), and Cantor (1872).

Though mathematical logic lost in this decade two of its great pioneers, Boole and De Morgan, it nevertheless flourished, for the legacy of Boole and De Morgan was taken up in the 1860's by Jevons, C. S. Peirce, and others. And the name Jevons may serve to remind us that the dream of a mechanical calculator, which was the legacy of Babbage (who died in 1871), was not then forgotten.

THE PHYSICAL SCIENCES IN THE 1860's

Many of the developments in physical science in the 1860's cluster around four central ideas. These are the ideas of a physical field, of energy, of the atom, and of spectrum analysis.

The discovery of the dark line spectrum occurred early in the nineteenth century; however, the true beginning of spectrum analysis is associated with the late 1859 publication of Bunsen and Kirchhoff. These authors explained the relationship between dark and bright line spectra and the relationships of these spectra to chemical composition. Thus as the 1860's began, scientists were given a new instrument of great power, and they were not slow to make use of it. By 1863 four new chemical elements had been detected spectroscopically, and in 1866 helium was detected for the first time, being found, not on earth, but in the sun. Moreover, in 1864 William Huggins solved what he called the "riddle of the nebulae," finding that certain nebulae give bright line spectra, and hence are not clusters of stars but rather luminous gases. Huggins went on in 1867 to make use of Doppler's principle and the spectroscope to measure the radial velocity of stars. Secchi's spectroscopic classification of stars made in the 1860's may serve as a last example of the great value of the spectroscope for the astronomer. In 1850 the astronomer had one instrument, the telescope; by 1870 he had three, for not only had he acquired the spectroscope during this time but he had also learned to take advantage of the advancements in photographic methods. Physicists likewise benefited from the spectroscope as they began to collect the information that would later play such an important part in atomic theory; fundamental to this was the spectroscopic map published in 1868 by Ångström.

In the history of atomic theory in the centuries that separate the atom of Democritus from the atom of Bohr, two names and two dates stand out. The first name is that of John Dalton who in 1808 published his *New System of Chemical Philosophy*. The second name is not that of a man, but of a town—Karlsruhe—and the date associated with it is 1860.

Dalton's book marks the beginning of modern atomic theory, but only the beginning, for when Dalton died in 1844, his ideas were neither universally accepted nor entirely acceptable. One chief problem was a confusion in atomic weights. In 1860, the first international chemical congress was held at Karlsruhe with some 140 prominent chemists in attendance. At that meeting Stanislaus Cannizzaro read and distributed copies of a paper in which he showed how, by means of accepting Avogadro's hypothesis of 1811 as essentially correct, chemistry could be unified and placed on a solid theoretical foundation. The reform suggested by Cannizzaro was, in fact, widely accepted. Lother Meyer stated that when he read Cannizzaro's paper, "The scales seemed to fall from my eyes." Symbolic of the fruitfulness of this reform for general chemistry is the fact that late in the 1860's Meyer and the Russian Mendeleev (who had also attended that congress) simultaneously announced the periodic table of elements. Symbolic of the fruitfulness of this reform for organic chemistry is the work of Kekulé, a leading figure in organizing the Karlsruhe Congress, who in 1858 had introduced the idea of the quadrivalency of carbon and suggested that carbon atoms could link with other carbon atoms. In 1865 he introduced the benzene ring structure. Kekulé's ideas combined with the acceptance of Cannizzaro's reform made the idea of valence acceptable, and moreover they mark the beginning of the study of "chemical structure," a term introduced in 1861 by Butlerov.

Thus in 1861 Kekulé had defined organic chemistry as the chemistry of carbon compounds. This definition, combined with Berthelot's famous book of 1860, *Organic Chem-*

istry Based on Synthesis, marks the final expulsion of vitalistic ideas from organic chemistry.

Such ideas and successes as those of Kekulé, Butlerov, Meyer, and Mendeleev strengthened atomic theory, and though some objections remained (as is shown by the debate in 1869 at the London Chemical Society), most chemists were convinced of the importance and usefulness of the atomic theory.

The physicists of the second half of the nineteenth century viewed the atom with both greater openness and more confidence than the chemists. The popularity of Kelvin's vortex atom illustrates this openness, and the rapid development of the kinetic theory of gases reveals the physicists' confidence in an atomic view of matter.

In 1858 Helmholtz had published a famous paper on the hydrodynamics of vortex motion in which he showed that vortex rings in an infinite, frictionless fluid were, under certain conditions, stable. In 1867, William Thomson (later Lord Kelvin), stimulated by the Helmholtz paper, propounded the idea that atoms were vortex rings in the aether. He and others were developing this theory until around 1900 when it became unacceptable. This now forgotten idea, however, was long viewed as having great promise. Thus, John Theodore Merz, writing at the beginning of the twentieth century, referred to it as "the most advanced conception in this line of thought, of which the human mind has so far been capable. . . ."[2] For our purposes, the widespread interest in this theory is especially interesting as revealing the inclination of many nineteenth-century scientists toward kinetic explanations and likewise their freedom from the billiard ball mechanics so often associated with nineteenth-century physics.

Before the 1850's few physicists were interested in atomic ideas; this situation was to change, for in 1857 Clausius brought forth the first major publication in the kinetic theory of gases. Maxwell's famous papers began soon after this, and by 1870 the kinetic theory of gases had become an established branch of physics. Though the work of Maxwell and Clausius was of the greatest significance, others also participated, and of these space allows only the mention of Loschmidt who in 1865 established the number of molecules in a given volume of gas, i.e., Loschmidt's number. One noteworthy aspect of the development of kinetic theory is that in the development of this subject statistical considerations for the first time played a major role in physical science.

We may now turn to the concept of energy and its status in the 1860's. It was in the 1840's that the first law of thermodynamics was discovered and developed by such scientists as Joule, Helmholtz, Mayer, and Colding. The second law of thermodynamics appeared in the early 1850's and is generally credited to Lord Kelvin and Rudolf Clausius. It is worthy of note that it was in 1865 that Clausius presented and named the entropy function.

With the development of these laws, scientists were set the task of re-examining many theories and phenomena in terms of the energy and entropy concepts. During the 1860's these concepts cast light in many directions; they were used to advantage in such diverse investigations as the study of the source of solar heat and Wunderlich's studies on the relation of fever to disease. Old textbooks were rewritten in such a way as to exhibit energy principles; new books likewise stressed it. Tyndall's *Heat as Mode of Motion* appeared in 1863, and the imponderable caloric, which had been favored even as late as the 1856 *Encyclopedia Britannica* article on "Heat," became a thing of the past. Typical of the stress on energy considerations in mechanics is Kelvin's and Tait's *Treatise on Natural Philosophy* of 1867 which was often compared to Newton's *Principia*. Energy considerations aided Maxwell in his electrical investigations, and, perhaps most importantly, the whole field of physical chemistry began to open up. Though the systematic formulation

Science a Century Ago

of physical chemistry came after the 1860's, the stage was set by such work as that of Deville on dissociation, Berthelot and Sainte-Gilles on reaction rates and heats of reaction, Guldberg and Waage who in 1864 announced the law of mass action, Traube who introduced semi-permeable membranes in 1867, Andrews on the liquification of gases and the development of the concept of critical temperature (1869), and finally Horstmann who in 1869 applied the concept of entropy to sublimation phenomena.

The fourth major development in physical science in the nineteenth century was the concept of a physical field. The idea of a substance that fills all space but that is not directly detectable by our senses dates back to the Greeks. Among seventeenth-century scientists, Descartes is only the best known of those who made use of the concept of an aether. By 1800 aethereal media had been used in a number of ways by various scientists to account for the transmission of gravitational, magnetic, electric, and luminous effects. However, in 1800 the majority of physicists accepted action at a distance for gravitational, electrical, and magnetic effects, and the then dominant corpuscular theory of light relieved them of the need for an aether for the transmission of light. The development of the mathematical theory of fluid motion (hydrodynamics), which may be traced back at least to Euler, and the development of potential theory, which is associated with Lagrange, Laplace, Green, Gauss, and others, served to supply some of the mathematical techniques that were being used when the situation changed.

And indeed the situation did change when in the first quarter of the nineteenth century the corpuscular theory was overthrown through the combined assault of Young and Fresnel, who in establishing the wave theory likewise established the need for a luminiferous aether. Certain polarization phenomena required that the waves be transverse rather than longitudinal (like sound waves), and this in turn required that the aether have the rigidity of a solid, yet be sufficiently soft, to allow the planets to move through it. Fresnel, Cauchy, MacCullagh, and Stokes were among those who before 1860 endeavored to write the equations of such an aether.

In addition to these developments, Michael Faraday, who was unable to accept action at a distance, began in the middle third of the century to introduce his concepts of lines and tubes of magnetic and electrical force. These lines and tubes implied that electrical and magnetic stresses were located in space. Faraday's ideas gained support from a series of papers written by Kelvin, the first of which appeared in 1842. Herein Kelvin called attention to mathematical similarities between equations of heat flow, current flow, elasticity, and certain equations in electrostatics and electrodynamics. The fact that equations used for heat flow in material substances are identical to equations describing electrostatic forces in space suggested that it was not unlikely that space, rather than being empty, is filled with an electrical aether. Maxwell commented that in this way Kelvin had "introduced into mathematical science that idea of electrical action carried on by means of a continuous medium, which, though it had been announced by Faraday . . . had never been appreciated by other men of science. . . ."[3] In the late 1850's the young James Clerk Maxwell, stimulated by Faraday and Kelvin, published the first of those four papers which were so important in nineteenth-century electromagnetic theory. The fourth and culminating paper in this series appeared in 1865, and in it Maxwell set forth the idea of a a displacement current, the electromagnetic theory of light, and the classic equations which bear his name. Thus light was interpreted as an electromagnetic wave, and electric and optical phenomena were referred to a single aether. In this last paper, Maxwell set aside the aether which he had developed in his earlier papers with the comment that it was heuristic and illustrative; however, he did comment: "In speaking of the Energy of the field . . . I wish to be understood literally."[4] That Maxwell was nonetheless convinced of the necessity of an aether is revealed by his article "Aether" for the *Encyclopedia Britannica*. Most physicists shared this belief, which was eventually destroyed, and with it the vortex atom, by the theory of relativity. It is well known that the theory of relativity gained major support from its ability to account for the Michelson-Morley experiment of 1887 in which the question of the earth's motion through the aether was investigated. It is less well known that in the 1860's Resphigi, Hoek, Maxwell, and Fizeau published experiments that were similar to the Michelson-Morley experiment in their aim, though not in their success. Though the aether has vanished, the associated concept of a field proved to be of a hardier stock, and the importance of Maxwell's famous equations is indicated by the following remark made by Einstein and Infeld in their book *The Evolution of Physics:* "The formulation of these equations is the most important event in physics since Newton's time, not only because of their wealth of content but also because they form a pattern for a new type of law."[5]

THE BIOLOGICAL SCIENCES IN THE 1860's

Slightly more than a month before the 1860's began, English book dealers were offered a new book published by John Murray of London. Earlier the author had written to Murray to caution him that the book might sell poorly. Murray nevertheless brought out an edition of 1250 copies, all of which, to the author's and Murray's surprise, were purchased by the book dealers on the first day of sales. On January 7, 1860, a second larger edition of the book was published; its title was *On the Origin of Species by Means of Natural Selection or the Preservation of Favoured Races in the Struggle for Life*. In 1872 Charles Darwin issued the sixth edition of the book and substantially revised it. Already by 1870 this book had necessitated a major revision of much of biological thought, for evolutionary theory had important implications for such diverse fields as taxonomy, ecology, comparative anatomy, embryology, and paleontology.

Some biologists accepted Darwin's ideas immediately and indeed devoted large portions of their lives to working out the ramifications of these ideas; others, however, rejected them. Symbolic of the controversy is that America's two leading biologists of the 1860's, Asa Gray and Louis Agassiz, took directly opposed views in regard to Darwin, much to the confusion and, I suspect, delight of their Harvard students.

Support for Darwin's ideas came from the brilliant enthusiast, Ernst Haeckel, who in 1866 elaborated his theory of recapitulation. "Ontogeny recapitulated Phylogeny," wrote Haeckel, and though he was neither entirely original nor entirely correct in this statement, nevertheless this "law" illuminated Haeckel's own researches as well as those of his associate Kowalewsky whose publications of the embryology of Amphioxux date from 1866.

In England Darwin received support from such eminent scientists as Hooker, Huxley, and Lyell. Though Darwin had declined discussion of the evolution of man in his *Origin of Species,* the topic was soon taken up by other authors. Lyell's *Geological Evidences for the Antiquity of Man* and Huxley's *Man's Place in Nature* both appeared in 1863, and in 1864 the co-discoverer of the theory of natural selection, Alfred Russell Wallace, published the first of his many discussions of this question. Finally, in 1871 Darwin's *Descent of Man* appeared. Just as Wallace found stimulus from and support for the evolutionary theory in his important investigation of the geographical distribution of animals, so also Darwin published during the 1860's important researches concerned with the interrelationships between insects and plants. The solitary Darwin received much support from the vocal Huxley, who defended Darwin's ideas not only against the Owen-coached Wilberforce but also against the more scientific criticisms of such men as Kölliker (1864).

Of the nineteenth-century objections to Darwin's thesis, perhaps the two most important date from the 1860's. In 1865 Lord Kelvin, well known to the public for his work on the transatlantic cable, published a paper entitled "The Doctrine of Uniformity in Geology Briefly Refuted." Herein Kelvin, arguing on the basis of physical measurements of the heat lost at the earth's crust, denied to biological and geological evolutionists the time span needed by them for the evolutionary processes. Though Kelvin's conclusions were wrong, as was seen after 1900 when subterranean radioactive sources were discovered, nonetheless Kelvin, speaking with the prestige of the physicist, had presented, as Darwin said, an "odious spectre" for evolutionary ideas.

Strong objections to the Darwinian theory also came on genetic grounds. Symbolic of these is an article published in 1867 by Fleeming Jenkin, who argued that chance variations in individuals would be "swamped" as the individual bred with other members of the species. Partly because of such problems, Darwin put forth his theory of Pangenesis in his 1868 book, *The Variation of Animals and Plants under Domestication,* and also introduced the idea of the inheritance of acquired characteristics into the later editions of the *Origin of Species.*

As many authors have affirmed, Darwin's work marks an important shift in man's view of nature and biological science. Thus in the Darwinian period, man came to see nature in historical, progressive, and dynamic terms.

As was mentioned, the acceptance of Darwin's theory was hindered by lack of information concerning the forms and mechanisms of inheritance. Darwin's acceptance of Pangenesis indicates the primitive state of both observation and theory in this regard. However, around 1857, an Augustinian monk had begun experiments in pea plants in the garden of his monastery at Brünn in Moravia. In 1865 this monk presented to the Brünn Society for the Study of Natural Science the conclusions he had reached on the basis of these experiments. With this paper, Gregor Mendel laid the foundation for the science of genetics. Though some of Mendel's conclusions had been reached before, no one had unified them as did Mendel, nor had anyone presented the abundance of quantified evidence which Mendel gathered. Despite the fact that interest in genetics was widespread at that time, as is witnessed by Naudin's work in the early 1860's and Galton's publication of his book, *Hereditary Genius* in 1869, Mendel's work failed to attract attention until his ideas were rediscovered at the end of the century. It is noteworthy that both Galton and Mendel were proponents of a statistical approach to biological phenomena; indeed it has been stated that Galton's greatest contribution was his advocacy of exact measurement and mathematical calculation in regard to biological science. What Mendel had done was to establish the form of hereditary continuity and, consequently, to delineate an essential element in the form of biological discontinuity or evolutionary change.

Though Robert Hooke had observed cells in the 1660's, cell theory is usually dated from the 1830's when Robert Brown detected the cell nucleus and Schleiden and Schwann published treatments of cell theory. There is merit in this view, but it should not be forgotten that there were serious defects in the cellular ideas proprounded by Schleiden and Schwann. Their structural ideas were rather incomplete, and their view of development was erroneous; thus, for example, Schwann held that new cells arose by a process analogous to crystallization in a structureless substance within or even just outside pre-existing cells. Cell division had been witnessed in some cases before Schwann and quite frequently in the 1840's and 1850's, but an accurate description of the process came only in the 1870's and 1880's. In the interim a number of significant events occurred, of which the most important is the publication in 1858 of Virchow's *Cellular*

Pathology. Herein Virchow argued, more convincingly than anyone before, that the cell is the unit of life and that, as he stated, "Where a cell arises, there a cell must have previously existed." Moreover, Virchow argued that pathology should be approached from the cellular point of view. And what Virchow's book did for cell theory in regard to pathology, Kölliker's book of 1861 did for cell theory in regard to embryology. The major defect in pre-1860 works relating to cells was corrected in 1861 by Max Schultze, who argued that cells should be viewed, not as "vesicular structures with membrane, nucleus and contents" but as "a lump of protoplasm inside of which lies a nucleus." Schultze thus shifted the emphasis to the protoplasm and away from the cell membrane, which, he showed, need not be present and which, when present, hindered cell division.

In the same year as Schultze's publication, Gegenbaur published a paper in which he showed that the eggs of vertebrates are single cells, and this, when combined with Schultze's paper, made it clear that protoplasm is the link between generations. In 1862 Brücke published a paper in which he stated that the importance of protoplasm demanded that it should not be viewed simply as an unorganized fluid. Thus, these three papers established the concept of protoplasm which, according to one historian of biology, J. Walter Wilson, is "of more far reaching significance" for biology than is the theory of evolution.[6] New techniques for the study of cells assured the further advance of cell theory. Many of these techniques were introduced during the 1860's; the first half of this decade saw the introduction of analine stains and hemotoxylin, and the latter half brought paraffin embedding and the microtome.

"One cannot expect the doctrine of spontaneous generation to be abandoned as long as one serious argument can be presented in its favor . . . such a doctrine can be compared with the mythical monster which had many ceaselessly regenerated heads. They must all be destroyed."[7] So wrote

Louis Pasteur in 1860 when he was in the midst of his Herculean labors against spontaneous generation. Before Pasteur, the researches of such scientists as Redi and Spallanzani had greatly narrowed the question of spontaneous generation; indeed when Pasteur began his experiments, the specific question concerned the origin of certain microbes that appeared in fermentation. In 1861 Pasteur published a series of brilliant and direct experiments which refuted the arguments of those who opposed him. Also in this paper he announced his discovery of anaerobic organism and in 1864 presented the technique now called "Pasteurization." In this decade he also determined the organism that produces pebrine disease in silkworms and saw his ideas taken up by Joseph Lister, whose famous publications on antisepsis and the germ theory of disease date from 1867. Pasteur was not alone in his work in bacteriology, for in the 1860's Villemin had done important work relating to the organism that causes tuberculosis, and Davaine had worked on anthrax. Thus, the theoretical foundation of the sciences of microbiology and bacteriology was established, and the stage was set for the later work of Koch, Cohn, Tyndall, and Pasteur himself.

The two greatest physiologists of the first half of the nineteenth century were François Magendie and Johannes Müller; Magendie died in 1855 and Müller in 1858. When they died, the work begun by them did not cease, for included in the scientific legacy of each of these men was a student who was eventually to rival the greatness of his master. Magendie gave us Claude Bernard; Müller gave us Hermann von Helmholtz. Both Helmholtz and Bernard had made important discoveries before the 1860's, but it was in this decade that they published three great physiological treatises. In 1863 Helmholtz published his treatise on physiological acoustics, and in 1866 the third and final part of his treatise on physiological optics appeared. The discoveries contained in these books are too numerous to mention; their importance is indicated by the fact that they are still in use.

In 1865 Bernard published his *Introduction to the Study of Experimental Medicine,* that work in which Bernard stated his conviction that physics and chemistry should be applied in physiology and that, like physics and chemistry, physiology is truly a science. The significance of Bernard's book may be best understood by considering the following. In 1847 Bernard, substituting for Magendie in a lecture, told his students: "The scientific medicine which it is my duty to teach you does not exist. The only thing to do is to lay the foundation upon which future generations may build, to create the physiology upon which this science may be established."[8] Such was Bernard's hope in 1847; his book of 1865 indicates that by that year his hope had been fulfilled. By 1865 medicine had become a science, and Bernard's book marks this transformation which his earlier discoveries did much to bring about.

Bernard and Helmholtz were not of course the only important physiologists of the 1860's; Ludwig, Sechenov, and Brown-Séquard were making important contributions to animal physiology, as were Boussingault, Woronin, and Sachs to plant physiology.

CONCLUSION

When I began to prepare this study, I resolved to attempt to determine whether the discoveries of the 1860's played any decisive role in the development of science in the last three centuries. This was done partly as a defensive measure, aimed at avoiding the tendency to attribute naively and subconsciously too much significance to this decade.

As I read various historians of science, I found that this decade has at times been referred to in such terms as "a time of maturation," "a watershed era," or even as "a period of revolution." For example, the noted historian of biology,

Charles Singer, wrote: "Thus the modern period of biology may be said to open in our era about 1860. . . . The whole outlook on the nature of living things underwent a complete and profound change in the period of about twenty years following 1860."[9] And Walter F. Cannon recently made the following statement in regard to physics: ". . . the massive watershed between 'Newtonian' and 'modern' is not around 1900 . . . the great discoveries are not such things as X-rays but such things as the continuous spectrum. Even now it is easiest to explain relativity as the decision to choose Maxwell over Newton whenever a conflict between the two developed. Likewise current quantum ideas can be seen as a decision to live with *both* Faraday's fields *and* Faraday's electrochemical discontinuities. Finally, given Leibniz, Babbage, Boole, and De Morgan, the computer age follows with the addition of the appropriate instrument-makers."[10]

A careful reading of Einstein and Infeld's *Evolution of Physics* can only lead to the conclusion that the thesis of this book is that the modern period in physics should be dated not from the beginning of this century but from around the 1860's. Statements referring in special terms to the period around 1860 in chemistry have been made by Aaron J. Ihde in his recent history of chemistry,[11] but in no case have I found a systematic discussion of whether the 1860's (or any decade in the nineteenth century) was a period of special significance in the development of science. There does, however, seem to be substantial agreement among historians that the 1860's were crucial in the development of biology; thus for example Dawes has referred to Singer's view with apparent approbation,[12] and a recent paper by J. Walter Wilson which argued a similar thesis was commented upon favorably by Shryock and Zirkle.[13] Finally, Jacques Barzun and Gertrude Himmelfarb in studies centering on Darwin have used the term "revolution."[14] A systematic discussion of these points is beyond the present study; nevertheless, one historiographic comment may be in order.[15]

I wish to suggest that, just as political historians have studied periods of war to the neglect of periods of peace, historians of science have stressed revolutions to the extent of neglecting periods and events that may in some cases be of greater importance in the development of science. The period in the chemistry from 1770 to 1820 will supply a specific example. Two great chemists stand out from this period: they are of course Lavoisier and Dalton. The former did battle with the phlogiston theory, finally overthrowing it and replacing it by the oxygen theory of combustion; the latter introduced modern atomic theory. The historians who write of Lavoisier may use such awe-inspiring terms as "overthrow" and "revolution," whereas he who writes of Dalton can avail himself only of such weak terms as "rise of" and "introduction of." Despite this fact, it is certainly possible to argue that Dalton's influence was greater than Lavoisier's. The Lavoisier-Dalton instance is but one illustration of what I take to be a serious defect in our historiographic vocabulary and conceptualization. As one who has concentrated on the history of mathematics, I feel particularly oppressed by the fact that such grand-sounding words as "revolution" are not and should not be used to describe epoch-making events in the history of mathematics.

Thus, I have been led to suggest that we need new terms that will allow us to do justice to such men as Dalton and such fields as mathematics. After some months of sifting, the terms "formational" and "transformational" seemed least open to objections.[16] The adoption of these terms may allow the historian of science to describe historical periods with greater precision and to set these periods within the mosaic of history with less distortion of the whole.

In a transformational event, an accepted theory is overthrown by another theory, which may be old or new. In such an event, there is a struggle in which both sides more or less understand each other, but still sharply disagree. At the conclusion of the event, an area of science has been transformed.

In a formational event, an area of science is not transformed, but is *formed*. The discovery or theory that produces this effect is usually new, and by definition overthrows and replaces nothing. There may be opponents, but these opponents either do not understand it or deny its significance. It may be that a transformation is eventually produced, but this effect is an incidental and conceptually distinguishable result. Usually a non-theory[17] such as "sunlight gives no chemical information" or "there are no geometrics but Euclid's" is violated, but this is hardly revolutionary.

Finally I would suggest that if the above analysis is applied to the 1860's, it will be found that this decade was primarily formational rather than transformational (or revolutionary), for nearly all the discoveries that have been discussed are characterized by the fact that they replace nothing at all; they fill what had previously been a void. I would also suggest that a large number of the new discoveries were formational precisely in the sense that they brought about syntheses. The most striking aspect of such new ideas as those concerning evolution, genetics, protoplasm and the cell, field, energy, spectrum analysis, the atom and the periodic table, and algebraic structure is that they unify areas of science that had previously been separate.

NOTES

1. The historian of science is seldom confronted with the task of surveying a given decade. Various aids are available that can lighten this task, and chief among these are the various chronological tables of discoveries. Of special value are the following which either are chronological tables or contain such tables:
    a. Ludwig Darmstaedter, *Handbuch zur Geschichte der Naturwissenschaften* (Berlin, 1908).

b. Agnes M. Clerke, *A Popular History of Astronomy during the Nineteenth Century* (London, 1893).
  c. Felix Auerbach, *Geschichtstafeln der Physik* (Leipzig, 1910).
  d. Paul F. Schurmann, *Historia de la Fisica,* 2 vols. (Buenos Aires, n.d.).
  e. Herbert S. Klickstein, *Outline of the History of Chemistry* (table published by Mallinckrodt Chemical Works).
  f. Gordon Rattray Taylor, *The Science of Life* (New York, 1963).
  g. Fielding H. Garrison, *An Introduction to the History of Medicine* (Philadelphia, 1929).
  h. Kark E. Rothschuh, *Entwicklungsgeschichte Physiologischer Probleme in Tabellenform* (Munich, Berlin, 1952).
  i. Edward W. Byron, *The Progress of Invention in the Nineteenth Century* (New York, 1901).

The above works helped at the beginning, but the task soon became one of selection. The authors who aided in this regard are too numerous to mention.

2. John Theodore Merz, *A History of European Thought in the Nineteenth Century,* II (London, 1928) 57. This work is an extremely valuable source of information on what the scientists of the nineteenth century thought and also on how they viewed what they thought (as above).

3. As quoted without reference in Sir Edmund Whittaker, *A History of The Theories of Aether and Electricity,* I (London, 1958) 241–242.

4. James Clerk Maxwell, *Scientific Papers,* I (Paris, n.d.) 564.

5. Albert Einstein and Leopold Infeld, *The Evolution of Physics* (New York, 1961) 143.

6. J. Walter Wilson, "Biology Attains Maturity in the Nineteenth Century," *Critical Problems in the History of Science,* ed. Marshall Clagett (Madison, 1962) 15.

7. As quoted without reference in Herbert A. Lechevalier and Morris Solotorovsky, *Three Centuries of Microbiology* (New York, 1965) 35–36.

8. As quoted in J. M. D. Olmsted, *Claude Bernard: Physiologist* (New York, 1938) 41.

9. Charles Singer, "Biology," *Encyclopedia Britannica* III (New York, 1937) 617.

10. Walter F. Cannon, "History in Depth: The Early Victorian Period," *History of Science* 3 (1964) 34.

11. Aaron J. Ihde, *The Development of Modern Chemistry* (New York, 1964) 55, 257–258.

12. Ben Dawes, *A Hundred Years of Biology* (London, 1952) 55.

13. J. Walter Wilson, "Biology Attains Maturity in the Nineteenth Century," *Critical Problems in the History of Science,* ed. Marshall Clagett (Madison, 1962) 401–418. For the comments of Shryock and Zirkle, see pp. 447–466.

14. Jaques Barzun, *Darwin, Marx, Wagner* (Garden City, New York, 1958), discussed the "Biological Revolution" stemming from Darwin, and Gertrude Himmelfarb's position is clear from the title of her book: *Darwin and the Darwinian Revolution* (Garden City, New York, 1959).

15. The historian of science who undertakes this task shall have to deal with the interesting study made by T. J. Rainoff, entitled "Wave-like Fluctuations of Creative Productivity in the Development of West-European Physics in the Eighteenth and Nineteenth Centuries," *Isis* 12 (1929) 287–319. Rainoff argues, on the basis of graphs in which numbers of discoveries are plotted against time, that the period 1850 to 1870 was a period of relative decline in productivity in physics.

16. The terms "mutational" and "transmutational," borrowed from biology and chemistry respectively, have some merit. It is scarcely necessary to point out that I am not so sanguine as to believe that these terms will be adopted. The terms are but labels used in making a distinction, and, if I have hopes, it is for the acceptance of the distinction. Though this distinction may be read as a criticism of Thomas Kuhn's ideas on scientific revolutions, it is the present author's intent that it may serve to *supplement* in a small way Kuhn's brilliant analysis.

17. It may be helpful to specify what I mean by a non-theory. A non-theory is a negative statement, usually held only implicitly, which is so general that no empirical tests are directly suggested by it. By way of qualification, not all negative statements are non-theories; Pasteur became famous for his efforts toward proving a negative statement which had some empirically testable implications. Non-theories are usually but not necessarily implicit; this is rather obvious since it is inconceivable that scientists would frequently make statements of the form: "I have no tests and can imagine none, but am convinced that X does not exist."

Prospects
and Challenges

5

CONTEMPORARY SCIENCE AND PHILOSOPHY

Herbert Feigl

As I understand the rather ambitious task assigned to me in this Symposium, I am to discuss the relations between science and philosophy in the scene of contemporary scholarship and even to cast a forward glance toward their future development. It should be obvious that such an undertaking is precarious. It would be difficult for anyone to be completely objective and impartial in matters so highly controversial. Moreover, if one were able to forecast the future of the scientific and philosophical outlook in any degree of specific detail, this outlook would already be present and hence no longer be in the future. This is itself the conclusion of a simple philosophical reflection on the predictability of intellectual achievement. Since I am not a visionary, it might seem necessary to restrict my presentation to generalities and platitudes. And since philosophers are in a way specialists in generalities, I do intend to deal with pervasive and basic issues but hope to avoid the platitudes by concentrating on several points of genuine relevance to the current controversies regarding the role of philosophy in our age of science.

Unavoidably my limitations of competence, as well as my personal predilections, will be reflected in my choice of topics and my way of discussing them. Furthermore, space limitation forces me to formulate matters in a tone more dogmatic than I would like, but for the sake of brevity and intelligibility I see no other way out.

In the following series of observations I wish to point out a certain very hopeful convergence, highly characteristic of what I consider a most fruitful conception of the task of philosophy. It requires no elaborate argument to show that ours is a new age of science. The practical results of the applied sciences surround us almost everywhere in contemporary civilization. There are immense problems regarding the proper utilization of the modern technological achievements. Consider the profound revolution in our ways of living, for example, through the uses of nuclear energy and of automation; the imminent application of biological engineering; and the new techniques of socioeconomic planning. How all this will affect the future of human welfare (involving the poignant issues of international peace and justice) is a topic that any reflective person will wish to ponder with whatever wisdom he can muster. Special attention is also being given nowadays to the "cleavage" in our culture between the outlook in the sciences and that of the humanities. This cleavage is manifest not only in the difficulties of communication between the representatives of the two orientations but also is central in the disputes about the ideals of education in our age of science.

It is precisely in connection with these issues that philosophical reflection can be particularly helpful. The vast and imposing structure of modern science, the tremendous variety of its methods of observation, experimentation, statistical designs, and theory-construction are being illuminated by philosophical analysis. The interest is focused here on the *pure,* rather than the applied sciences.

Under the impact of modern science, the very conception

of the task of philosophy has undergone a radical reorientation. In very broad historical perspective it may be said that the philosophers throughout the ages have been engaged in three major endeavors: they have been searching for absolute truths regarding basic reality and for absolute standards of morality; they have tried to construct a synthesis, a view of the universe and of man's place in it, i.e., a perspective that would integrate the various contributions of the sciences into an intelligible and harmonious whole. Finally, philosophers have attempted to clarify the meaning and the validity of the fundamental concepts, assumptions, and methods of knowledge and of evaluation.

In the light of modern science, i.e., not only of its results and conclusions but especially of its open-minded attitude and critical approach, the search for absolute truth is largely being abandoned as fruitless, if not as meaningless. The spirit of contemporary science is "critical" in the sense that all its conclusions are considered sound or tenable only "until further notice." In other words, neither of the extremes —dogmatism or skepticism—is acceptable. The policy of the open mind indicates that while we should keep all our convictions in principle open for criticism and revision, it is perfectly reasonable to rely on well-confirmed assumptions until strong evidence forces us to modify or to replace them by other assumptions that are more strongly supported by relevant evidence.

The search for the Absolute has become suspect because assumptions regarding it are usually made proof against disproof, that is, they are not susceptible to any sort of test. Nevertheless, at least psychologically speaking, the "will to believe" is strong and expresses itself philosophically in various forms of metaphysics and theology. From a scientific point of view it must be asked whether the truth-claims of transcendent, i.e., radically transempirical, beliefs are not illusory in that they mistake basic moral commitments for genuine knowledge claims. Logical analysis of these issues impels us to distinguish between scientific and non-scientific endeavors. There is no need to be *"scientistic,"* i.e., to use the term "nonscientific" in a disparaging manner. The very aims and hence the criteria of appraisal of the arts, of literature, or of music are different from those of the sciences. These *non*scientific activities must not be confused with *un*scientific pursuits. We reject *un*scientific endeavors (e.g., astrology), for although their *aims* are those of science, namely, explanation and prediction, the procedures used are discredited in the light of the criteria of good scientific method.

The search for the Absolute today is best interpreted as a matter of basic moral commitment rather than as a guess at the riddles of the universe. As the (transempirical) metaphysicians have conceived and treated these riddles, they have become absolutely unanswerable questions, "guaranteed 100 per cent insoluble!" In contrast to that, the sciences have construed at least some of these problems in ways that make them accessible to increasingly more adequate solutions. But if the sciences take over this task, is there anything left for philosophy to do? It must indeed be admitted that the grand-style syntheses that were still so fashionable early this century have become highly questionable. In earlier epochs the philosophers, if they were thoroughly conversant with the sciences of their time, could still suggest new "world-hypotheses" and in this way blaze the trail for further advances of the sciences. Vitally important and deeply fascinating as is the task of the integration of a world view, in our age it requires a high degree of scientific competence. This the vast majority of philosophers regrettably do not possess. Moreover, it is through the autonomous development of new borderland *scientific* disciplines that genuine progress towards a justifiable integration is being achieved.

Speculative anticipations of scientific syntheses and unifications can hardly be regarded any longer as a proper task for philosophy. It is true, there have been occasions in the past when thinkers who may be more properly designated as philosophers rather than as scientists have introduced novel and fruitful ideas which were later developed, and usually in strongly modified forms confirmed by scientific procedures. The atomic hypothesis of Leucippus and Democritus, the Ionian anticipations of evolutionary theory, Aristotle's views of organic life, the perceptive remarks of Nietzsche on the psychology of the unconscious—these and possibly just a few other instances are pertinent illustrations. But it seems that the imagination of the scientists in recent times has far outrun that of the philosophers and the poets. No philosopher has even vaguely anticipated the theory of relativity or the quantum theory. Nor have the theories of genetics or of biochemistry been foreshadowed by philosophical speculations. It does seem that today and in the future any sort of fruitful theorizing can be expected only from competent scientists. They must have an expert grasp of the observational and experimental facts, they must be thoroughly conversant with the extant theories, and they have to be able to utilize advanced mathemetical techniques. The great scientific syntheses of the modern age have often been prepared by the development of borderland sciences, such as the "hyphenated" disciplines of physical chemistry, biophysics, biochemistry, psychophysics, psychophysiology, social psychology, socioeconomics, and such auxiliary disciplines as cybernetics and information theory. To be sure, many philosophically-minded scientists have initiated, or contributed to, these developments.

I do not mean to say that professional philosophers of science are disinterested in synthesis or synopsis. Quite to the contrary, a good deal of their work concerns precisely the logic of unifying syntheses. They share the scientific theorist's fascination with the synoptic power of the hypothetico-deductive method. One may interpret this fascination psychologically in various ways. There is the "economic" aspect: theories are attempts to cover a maximum of facts by means of a minimum of basic concepts and assumptions. The predictive efficacy of scientific theories in a way demonstrates how much one can get for how little. Another aspect may be that of a sublimated power drive. Instead of dominating other persons, the scientific theorist masters and marshals a great number of facts. I refrain from discussing "deeper" psychoanalytic interpretations according to which the quest for scientific understanding may be traced back to infantile curiosities. No matter how we explain the fascination, it exists and it accounts for at least one aspect of the high regard for this monumental contribution of the human intellect, the theories of modern science. Hence it is understandable that philosophers have attempted to achieve a "science of science" which would make the almost miraculous achievements of scientific theories somehow comprehensible. Perhaps we should remember Einstein's famous saying: "The most incomprehensible thing about the world is that it is comprehensible!" The wish to *understand* the nature of science is perhaps the purest and most commendable motivation for philosophical analyses of science. But, of course, other motivations have also yielded worthwhile results. There are logicians who use the sciences as a domain in which to exercise their formal (and often formidable) intellectual powers. And there are others, perhaps more artistically inclined philosophers who prefer to rhapsodize quite informally about the scientific enterprise. Still others are preoccupied with epistemological or metaphysical puzzles and prefer to tackle them within the arena of "metascientific" reflections.

To summarize: What, then, is the proper task of philosophy in our age of science? The fairly generally accepted answer seems to be: *Analysis,* i.e., the third task I mentioned previously. This is often misunderstood to amount to nothing but destructive criticism. A closer look tells us, however, that analysis can be quite *constructive.* While it may well start with the elimination of needless perplexities by exposing the underlying conceptual confusions, analysis often does illuminate the very logic of syntheses and integrations. In the forefront of current philosophy of science are studies in the logic and methodology of scientific theories as well as the interrelation of the sciences.

It is one of the marks of the modern scientific temper in philosophy that two undesirable and diametrically opposite dangers are being avoided. These are typified by the philosophies of the *"nothing but"* and of the *"something more"* type. That means we are aware of the pitfalls of both the "reductive" and the "seductive" fallacies. The first reduces the world, or man's place in it, to an absurdity, as in "crass mechanistic materialism," and the second reads mysteries into the world for which there is no evidence. Only a philosophy of the *"what's what?"*—i.e., a genuinely *constructive* or *reconstructive* analysis—can do justice to the intricacies and complexities of the world.

There is a fair measure of agreement today on how to conceive of *philosophy* of science as contrasted with the history, the psychology, or the sociology of science. All these disciplines are *about* science, but they are "about" it in different ways. The history of science traces the development of scientific problems, ideas, and solutions preferably within the entire socio-cultural context. The psychology of scientific discovery tries to account for the creative, problem-solving activities of the scientists in terms of the requisite mental processes. The sociologist of science tries to account for the development and reception of scientific theories and points of view. Investigations of styles and fashions in scientific theorizing, reflecting the *"Zeitgeist"* of a given period of cultural, social, economic, or political conditions clearly belong in this area of the sociology of knowledge.

In the widely accepted terminology of H. Reichenbach[1] studies of this sort pertain to the *context of discovery,* whereas the analyses pursued by philosophers of science pertain to the *context of justification*. It is one thing to ask how we arrive at our scientific knowledge claims or what socio-cultural factors contribute to their acceptance or rejection; and it is quite another thing to ask what sorts of evidence and what general objective rules and standards govern the testing, the confirmation or disconfirmation, and the acceptance or rejection of knowledge-claims in science. The question "How do we know?" is thus typically ambiguous; it may amount to asking "How did we come to know?" or it may mean asking "What reasons can we give by way of objective support for our claims to know?" As a prerequisite for answering the second, i.e., the philosophical kind of questions, we need to be clear about the *meaning* of scientific assertions. This involves a scrutiny of the logical structure of scientific concepts, and it involves critical reflections upon the lines of demarcation of the scientific from non-scientific and unscientific endeavors. These two questions, then, "What do we mean?" (i.e., by the words and symbols we use) and "How do we know?" (i.e., to be true, or confirmed as likely to be what we claim as scientific truths) guide the inquiries of modern philosophy of science. Analyses of the empirical bases and of the logical structure of the factual sciences are the primary concerns. On many signal occasions great scientists have been their own philosophers in that they have produced new theories or methods aided by incisive critical scrutinies of the total relevant conceptual frame. But with the increasing specialization in all fields, at least two generations of philosophers of science have arisen. Competent understanding of at least some scientific disciplines, combined with logical acumen and philosophical insight, is a distinction achieved only by a rather small group of philosophers of science. But the efforts of these specialists have yielded results which are eminently enlightening.

The approaches and procedures used by individual philosophers of science differ widely. They range from rather informal studies to formidably formal reconstructions. Personally, I think there is a place for all of these. Indeed, there are dangers at the extreme ends. Fortunately, the British preoccupation with the analysis of ordinary language did not (and really could not properly) make much headway in the philosophy of science. It has not strongly influenced American philosophy of science. On the other hand, the "precisionists," i.e., the constructors of formal systems, have contributed a great deal, and there can be no doubt that some work of this kind has been outstandingly fruitful. In the spirit of Euclid's method as it has been tremendously improved and refined in modern mathematics, entire disciplines in the empirical sciences have been cast in the form of deductive systems. The danger here is that formalization and axiomatization becomes an end in itself. The ideals of mathematical compression and "elegance" have their place, but not necessarily in the logic of the empirical sciences. Most needed here are systems of maximally independent testability. Such systems enable us to see which postulates of a theory are supported by what empirical evidence and, hence, which postulates may need revising or supplanting if and when contrary evidence should emerge.

There are important and controversial issues in the logical reconstruction of scientific theories. These have occupied some of the best intellects of our time and will no doubt be intensively studied and discussed in the coming years. One of them is the vexing question as to exactly how to construe the relation of theories (in the empirical sciences) with the sort of evidence that is relevant for their confirmation or disconfirmation. The customary reconstruction is the one originally proposed by N. R. Campbell and H. Reichenbach, and with certain variations developed by Carnap, Hempel, Margenau, Northrop, Braithwaite, and others.[2] According to this view of the matter, we are to distinguish between the *theoretical* language of a given discipline (like thermodynamics, relativity, quantum theories, the gene theory in biology, etc.) and the *observation language*. The terms of the theoretical language are "implicitly defined" by postulates, and a subset of these terms (or of concepts explicitly defined on the basis of those "primitive terms") are then connected by "coordinating definitions" or "correspondence rules" with empirically or ("operationally") defined terms of the observation language. It is understood that this analysis is a reconstruction, that is, not intended to reflect the origin or development of scientific theories. It was presented as a device that should enable us to scrutinize separately, on the one hand, the logico-mathematical (or "purely formal") aspects of a theory in regard to the consistency of its postulates and the validity of the deductive derivations and, on the other hand, the theory's empirical content and thereby its observational confirmation (or disconfirmation). Limitations of space prevent me from discussing the exciting controversy between R. Carnap and K. R. Popper[3] regarding probability, confirmation, and corroboration. (The entire issue of the possibility of an inductive logic depends on the settlement of this controversy.)

I do want at least to touch on one line of radical criticisms of the reconstruction of scientific theory. I am referring to several of Paul K. Feyerabend's publications.[4] He very properly points out that the theorems deduced from the postulates of a given theory could not possibly be formulated in the language of direct observation. The simple reason for this is that if it is to be *deduction,* then the concepts appearing in the theorems must either be the primitives themselves or (more usually) concepts explicitly defined in terms of the primitives. But the terms of the observation language are clearly not thus definable. This was of course the consideration that led to the notion of the correspondence rules. In regard to these, Feyerabend insists that the very logic of the observation language does not mesh with that of the theoretical language. Indeed, he goes further and maintains that even the observation language is already suffused (or contaminated) with a theory, at least a crude one. Thus, to use a well-known example, the concept of temperature as used by the experimental physicist cannot be coordinated, let alone identified with the concept of the mean kinetic energy of the molecules. Why not? Feyerabend replies that the experimental-mensurational concept of temperature involves laws (i.e., at least a low-level theory) which are logically incompatible with the kinetic theory of heat as developed in the statistical mechanics of Gibbs and Boltzmann. It is true that the classical second law of ("phenomenological") thermodynamics turns out untenable in statistical mechanics. Feyerabend therefore proposes that philosophers of science should abandon the entire idea of the level structure of scientific explanation according to which the lower level theories and experimental laws were supposedly deducible from the postulates of higher level theories. In place of that familiar reconstruction, Feyerabend declares that we should merely retrace the successive steps in which earlier, less adequate theories are supplanted by later more adequate ones. This move, however, forces him to say that the testing of theories involves an observation language that is already formulated in terms of the concepts of the theory to be tested. This consequence seems to me discordant with the actual procedures of the confirmation or disconfirmation of theories. The very requirement of testability, the need to account clearly for what is involved in the testing of rival theories and deciding which among them is most strongly confirmed (or as Popper would have it, "corroborated," i.e., surviving strenuous attempts to refute it by observational evidence)— all this and the obvious claim that the latter and better theories which replace the earlier and poorer ones share (at least roughly) the same subject matter seem to me to point to a more conservative reconstruction.

Although the precise details still need to be worked out—and this is an intricate task—I submit the following formulation: the identity of the subject matter may be explicated by the approximate congruence of the regularities deduced from an advanced theory with the regularities determined by experimentation or just plain observation and their inductive interpolations and extrapolations. There is thus always at least a measure of agreement within a limited segment of the relevant variables. I mention as pertinent illustrations the practical certainty of entropy-increase in case of macroscopic processes; the quantitative agreement of classical mechanics and Einsteinian special relativity within the range of fairly low velocities; the "correspondence" (actually so called by Niels Bohr) between the experimental regularities derivable from classical electrodynamics and of the quantum theory. Of course, what interests us just as much as these correspondences are the (often tremendous) discrepancies outside a certain range of the pertinent variables. Surely, Feyerabend is right in stressing these when concerned with the progress and the revolutions of science. But I do not think that his radical philosophical conclusions follow from an analysis that takes due account of permanence as well as of change in the succession of scientific theories. What is called for is, of course, a re-examination of the logical nature of the correspondence rules. In agreement with Ernest Nagel,[5] I am inclined to view the correspondence rules, at least in one important form of their employment, as *bridge laws*. As such they differ sharply from mere semantic rules of designation; they relate, e.g., concepts of microentities with concepts of observable macro-entities. And once various lines of converging evidence are available for the micro-entities and their characteristics, these bridge laws stand or fall in the light of empirical evidence.

Although the preceding remarks may seem somewhat recondite, if not esoteric, I do think that they have a significant bearing on some of the most poignant philosophical issues in our age of science. Permit me therefore to utilize the observations thus far presented in a brief discussion of three questions that are of paramount importance in regard to the alleged cleavage in our culture, i.e., the relation of science and the humanities. The three problems I wish to tackle are, first, *The Limitations of Science;* second, *The Mind-Body Problem* (or the *Problem of Man's Place in the Universe*); and, closely connected with these, the third, *The Freewill Problem.* Any even half-way proper discussion of each of these tremendous and controversial issues would take several essays. But given an enlightened audience, I am confident that a few succinct remarks on each question will be intelligible and, I hope, fruitful.

Besides the demarcation of science from the nonscientific endeavors, various types of allegedly insuperable *limitations* of science have been widely discussed. Although the famous *"ignoramus et ignorabimus"* of E. du Bois-Reymond is now almost forgotten, and E. Haeckel's rejoinder in his naively metaphysical and dogmatic *The Riddles of the Universe* is safely buried amidst other Victorian fossils of thought, there are more recent and more serious considerations regarding the ultimate limitations of scientific explanation and prediction. K. Gödel's[6] proof of the essential incompletability of all postulate systems of mathematics (involving at least denumerable infinity) may well be relevant for the sort of mathematics needed for the derivation of theorems in modern theoretical physics. The status of the quantum mechanical uncertainty and the corresponding, possibly ineluctable statistical character of the basic laws regarding the atomic and subatomic processes appears to present us with an insurmountable difficulty in the search for a strictly deterministic world order. The logical clarifications of the sort of inferences involved in scientific explanation have made it abundantly clear that all such explanations are relative in two ways: *one,* that the *explanantia,* i.e., the law-like postulates which serve as premises depend for their acceptability on the always-in-principle-incomplete and indirect observational evidence; and, two, that these premises are themselves unexplained in the context of the given explanation. And although it has often been possible to rise to higher strata of explanations, there are always theoretical postulates which are themselves underived and must (at least "until further notice") be accepted as "brute facts" pertaining to the fundamental order of nature. The attempts to interpret these as logically necessary or else as synthetic *a priori* truths simply rest on conceptual confusions. (I am referring here to the in other ways brilliant and sophisticated ideas of H. Poincaré, A. S. Eddington, H. Weyl, and even to some of the more pythagorean inclinations in the later thought of Einstein.) Similarly, any talk of absolute presuppositions or of the truths of metaphysical insight or intuition usually comes down to sanctimoniously formulated verbal sedatives. (Too many philosophers have sold their birthright for a pot of message!)

Applying sober analyses, such as these, to the perennial perplexities of the problems of the mind-body relation (or of man's place in nature), it may be asked what the sciences and philosophy can do about the so-called homeless qualities. I am referring here to the time-honored questions regarding the supposedly "completely abstract" representation of the universe in modern physical theory. In other words, how can the admittedly powerful and exact account of the world given by the most advanced physical theories be related, or even squared, with the qualities which are immediately experienced in all their warm and colorful varieties. The currently renewed and lively occupation with this baffling question indicates that the long siege of repression of the mind-body problem is finally being overcome. Radical logical behaviorism and physicalism (let alone crass materialism) are being attacked by highly sophisticated philosophical analyses. If I interpret the signs of our time correctly, a new solution is in the making which is neither viciously reductive nor wishfully seductive. We can afford again to satirize the radical behaviorist (who maintains that thought is "nothing but" subvocal movements of the larynx): "he has made up his windpipe that he's got no mind!"

I had the privilege of an afternoon's intensive and unforgettable conversation with Albert Einstein at Princeton, one year before his death. Although we talked mostly about the logic and epistemology of modern physics, we also touched on the mind-body problem. In this connection, Einstein, considering the abstract nature of the four-dimensional (Minkowski) representation of particles and fields, said that for a complete description of all there is in the world we would have to attach to that description another one, indicating the spots of "internal illumination" (Einstein's metaphor for centers of conscious awareness). And he remarked in his characteristically unpolished way and with cheerful laughter: "Ohne diese rein subjektiven Erlebnisse wäre die Welt doch ein blosser Misthaufen!" (In rough translation: "Without these purely subjective immediate experiences, the world would be a mere pile of dirt!")

Thus encouraged I renewed my attempts towards a scientifically acceptable and logically defensible nonreductive solution of the mind-body problem. As I see it, the main task here is to find a formulation that reconciles the features of both *sentience* and *sapience* with the physicalistic account of biological, and especially neurophysiological, processes. During the last seven years as well as currently, the debate is becoming increasingly more pertinent and extensive.[7] Of course, there are many thinkers who try to understand human thought and experience on the basis of the computer analogy. No doubt the "robotologists" are making important and challenging contributions. (But, understandably, most of them propose radically physicalistic views.) There can be no question that cybernetics and information theory as applied to the neurophysiological mechanisms (involving negative feedback, reverberating circuits, the functions of the reticular formation, and of the pleasure and displeasure centers in the brain, etc.) are contributing excitingly new results to the *scientific* explanation of animal and human behavior. This has helped greatly in counteracting the exclusively peripheralist approach of some of the behavioristic psychologists. At last it is respectable again to inquire what goes on inside the "black box!" Hence such behavioral constructs as habit strength, memory trace, drive-intensity, etc., may then be identified (by the use of correspondence rules in the sense of bridge laws) with the central-process-concepts in the scheme of neurophysiology. But, in agreement with Einstein's dictum, I think that a further step is needed. If I may (mis-)use a French saying, "C'est le premier pas qui coûte!" Indeed this is the first, and perhaps also the last, philosophical step which is crucial. It is on this step that the current controversy is focused. Are we to attach mentalistic concepts referring to *sentience* by way of "nomological danglers" to the otherwise purely physicalistic conceptual system; are we to admit the notion of genuinely emergent qualities; or can we make sense of an empirically based identification of the event labeled and described in a twofold manner, on the one hand in neurophysiological (and ultimately microphysical) terms, and on the other in introspective mentalistic (or phenomenological) terms? All this concerns primarily *sentience*. In regard to *sapience* we may ask: Can we identify the mentalistic notion of *intentionality* (stressed by many philosophers ever since Franz Brentano introduced it as *the* basic feature of all mental acts) with the designation relation of pure semantics? Surely these and other related and poignant questions are going to be discussed for some time to come. My own inclination is to start with *the qualities of immediate experience as the subjective or private aspect of neurophysiological processes* as an *explicandum* and proceed by logical and semantical *explication* towards a solution of the mind-body problem. But on my grayer mornings I realize that one is apt to slip into shameful inconsistencies if one endeavors to do justice to *all* the bewildering facets of this intriguing problem of problems. Small wonder that Schopenhauer regarded it as the *"Weltknoten"* (the world knot). Indeed, it seems that all roads in philosophy lead to the mind-body problem. It has remained a most recalcitrant issue throughout the entire history of thought.

Permit me to mention briefly the one glimmer of light that I perceive in the appalling darkness: closer attention to the semantic and pragmatic functions of the egocentric terms in our language may help us towards the resolution of the basic perplexity. As is generally understood, the egocentric particulars, viz., the *Now,* the *Here* and the *I,* extremely useful though they are in common discourse, simply disappear in the intersubjective language of science. They are replaced in the intersubjective system by definite descriptions of moments in time, locations in space, and persons. The— I am tempted to say—"existentially poignant uniqueness" of the *I,* the *Here,* and the *Now* is lost in the (as it were "democratized") objective account of the world. The *I* thus becomes one person among others, the *Now* one moment among others, the *Here* one place among others. Well then (and I do not know whether some other philosopher or logician has thought about it in just this way) perhaps the predicates which designate immediately experienced qualities similarly disappear in the intersubjective account and are replaced by predicates that have only a locus in the abstract structural description of science. Concept formation and theory construction in the physical sciences are *structural* in the sense that by themselves they merely reflect the nomological network but do not explicitly represent the intrinsic qualities, whatever they may be. It is conceivable that the denizens of another planet have an entirely different repertory of qualities in their immediate experience; and if they had a utopian or millennially precise physics and physiology, they might be able to explain and predict, at least statistically, all phases of human (the earthlings') behavior. But they might nevertheless have no intuitive, empathetic understanding of the qualities of human sensations, feelings, emotions, or sentiments.

In short, the approach towards a synoptic solution of the

mind-body problem that seems to me to hold some promise is a new version of the identity theory, or of the twofold knowledge view. In earlier forms this sort of solution was proposed by some of the American and the German critical realists; it was adumbrated in the structuralistic accounts of physical concepts and theories by Henri Poincaré, A. S. Eddington, and more clearly worked out in the essays on "Form and Content" by Moritz Schlick.[8] Among more recent American philosophers, F. S. C. Northrop[9] and S. C. Pepper[10] should be mentioned as forceful proponents of a similar view. My colleague Grover Maxwell, in several as yet unpublished essays has put forth a well-articulated structural realism. The main danger—of which Schlick was aware but did not quite manage to avoid—is to render pure content absolutely ineffable. The qualities of immediate experience, far from being ineffable, *can* be conceptualized—but of course by virtue of their logical structure. Therefore one would be quite mistaken to identify the mental with the pure content of the experienced qualities. Only that which is strictly subjective or absolutely private about these qualities is ineffable, and that for the reason that we are here (in what B. Russell used to call "knowledge by acquaintance") confronted with what some metaphysicians refer to as *"pure being."* Since genuine knowledge is always propositional, the communicable and intersubjectively testable knowledge claims of phenomenological psychology concern the *structure* (i.e., the logical form) of those very contents. And if this structure is isomorphic with certain Gestalt (i.e., configurational) aspects of the neurophysiological processes in our brains, it seems plausible to say that it is the very same reality (the numerically identical reality) that is known on the one hand through introspective awareness, and on the other on the basis of the observations of behavior and/or of neurophysiological processes. There are still many difficulties of a logical sort to be overcome by careful semantical analyses; but—if I am to hazard a very risky prophecy—this view just sketched seems to me a good bet on the future development of the philosophy of mind.

It is perhaps more a matter for terminological decision rather than a profound philosophical issue as to whether we call the attribution of directly experienced and private qualia a metaphysical, i.e., transempirical, nonscientific inference, or extend the definition of science so as to include those peculiar and directly untestable inferences within the domain of scientific inference. The logical positivists, at least in the heyday of the radicalism of the Vienna Circle, considered the issue as meaningless or else, as in Carnap's later views[11] as a question merely pertaining to a choice of language. A remnant of the Viennese verificationism is, I think, still noticeable in the later work of Ludwig Wittgenstein and in some of his disciples.[12] In a different form, even Karl Popper, whom no one can accuse of verificationism, seems forced by his unrelenting repudiation of inductive inference to relegate the ascription of private experience to the domain of metaphysics, "good metaphysics" in this case. As suggested before, I don't think very much hinges on the question as to how we label these extreme forms of analogical or inductive inference.

What does matter, however, is how in the light of the suggested analysis of the mind-body problem we view the relation of the sciences to the arts and music and literature. I deliberately did not say "to the humanities" because I consider the humanities, e.g., the histories of art and literature, as well as literary and art criticism, as cognitive disciplines which endeavor by systematic study and scrupulous scrutiny to make knowledge claims regarding their chosen subject matters.

There is a real danger of highly misleading oversimplifications. Just as F. S. C. Northrop[13] did not quite succeed in his attempt to contrast as well as to reconcile occidental and oriental philosophies by stressing conceptual form in the thinking of the West, and intuitive content in that of the East, so it would be similarly hazardous to contrast the sciences and the arts in terms of structure and content. It seems to me that the creative aspects in the arts and in the sciences have much in common if viewed *psychologically.* But *logically,* i.e., in the light of the *norms of criticism,* they differ in that the order of primacy in regard to the arts is exactly the reverse of that in the sciences. Whereas in the sciences cognitive consistency and a high degree of confirmation are the primary criteria, and the aesthetic qualities of harmony, elegance, and symmetry are secondary, it is exactly the other way around for the arts. Surely, the qualities of immediate experience are of paramount importance in the appreciation of the arts, and this is especially so because even the structure of the work of art elicits special experiential qualities. In contrast to this, the data of immediate experience play the twofold role of serving on the one hand as the ultimate confirming (or disconfirming) evidence, and on the other hand they are the objects of knowledge in the limited domain of introspective or phenomenological psychology. This, aside from many important details, seems to me a more appropriate characterization of the similarities and the differences between the arts and the sciences than any rash and superficial declaration to the effect that science is an art and that the arts are a sort of science!

Finally, let me present just a few observations on an issue (really a corollary of the mind-body problem) that has been perennially a bone of contention among scientists and philosophers, namely, the problem of determinism and free will. The question whether the world is "at rock bottom" deterministic or indeterministic is unanswerable, and really irresponsible, since there can be no criterion for rock bottom. We would not know rock bottom even if we saw it! Metaphysical speculation here seems entirely inconclusive, if not absolutely futile. All that can be asserted is that ever since the statistical interpretation of quantum and wave mechanics by Max Born[14] in 1926 and the various attempts by Einstein, Bohm, de Broglie, Vigier, and others to restore to modern physics a strictly deterministic foundation have thus far not succeeded. The vast majority of physicists today are quite reconciled to the idea of basic statistical laws and hence to probabilistic explanation and prediction. Some, indeed, seem even to say—by way of a "sour grapes" claim —that they never desired a 100 per cent determinism. Using the locution "statistical causality," they even tend to blur the issue, if not to make us believe that the laws of quantum mechanics are deterministic in that the pivotal magnitude 'Ψ' in Schrodinger's partial differential equations functions the same as the variables of classical deterministic physics.[15] However, it was just the essential point of (the generally accepted) view of Max Born that 'Ψ' by itself is not a magnitude pertaining to the real events in space and time, but that $|\Psi|^2$ represents the probability of such events. In other words, the interactions between particles and between particles and fields—which according to current physics are the *basic* processes in the universe—are a matter of *statistical* laws.

It is astonishing, and I think rather distressing, that even some prominent thinkers have taken the indeterminism of modern physics as a basis for free will. It should have been fully clear, at least ever since the illuminating analysis of the free will problem by David Hume (in the eighteenth century), that this problem in its traditional form rests on a confusion of causal determinism with compulsion (coercion, constraint), and the corresponding confusion of free will with absolute chance. If we accept until further notice quantum mechanical magnitudes as at the "rock bottom" of nature, this would give us—by amplification at most— absolute chance events in human behavior. And this is something very different from the sort of free choice which is presupposed by moral responsibility. Quite to the contrary. As the late Dickinson S. Miller (writing under the pseudonym R. E. Hobart) once put it in the title of a remarkable essay, "Free will involves determinism, and is inconceivable without it."[16] It is highly regrettable that this fundamental insight remains so widely misunderstood. The kind of free choice we have clearly consists, negatively, in the absence of compulsion and, positively, in our capacity for choosing and acting in accordance with our basic personality and character. That our personality and character are in turn determined by our inherited constitution plus a variety of environmental influences does not in the least detract from such genuine degrees of freedom as we actually possess. We *are* the doers of our deeds because we are (at least in most of the normal life situations) essential links in the causal chains of the events that make up the history of mankind. We are responsible to the extent that we are responsive to the influences (educational, sociopolitical, legal, penal, etc.) that society has at its disposal for molding our attitudes. The important point of regret, remorse, and repentance is that it is pragmatically prospective and not merely retrospective as in the phrase "I wish I had not done it." It is, rather, the resolution firmly made to the effect that on the next occasion we shall act differently.

At this point inevitably the question arises as to whether the basic principles of morality can be derived from the scientifically confirmed propositions about the nature of man. Many (but not all) empiricists and positivists, ever since David Hume's incisive analyses, have answered this question in the negative: the moral "ought" is not deducible from the factual-empirical "is." Although I believe this conclusion to be ultimately correct, it seems to me somewhat misleading in its stark bluntness. To be sure, it is difficult to avoid *petiones principii* in the justification of the fundamental maxims of morality.[17] Scientifically oriented philosophers repudiate metaphysical and theological premises. But they have perhaps not sufficiently appreciated the truths of Aristotle's ethics. Unduly impressed by the ethical relativism of the anthropologists of the nineteenth century, they did not realize that while there is a pluralism and relativism of *mores* (i.e., folkways) there are nevertheless certain *common* and constant features in the *moral* ideals of most cultures, ancient and modern, oriental and occidental. It is, of course, deplorable that we do not fully conform with these basic moral norms, that indeed we have witnessed in our own time even some of the most horrible violations. But this does nothing to disprove that human conscience in the social context of the processes of living together has come to insist on the moral ideals of justice and fairness, kindness and love, as well as of self-perfection.

In view of the ever more rapid and extensive application of science to human affairs, it will be imperative to keep scrutinizing from a moral point of view the effects of automation, of socioeconomic and political planning, of the application of the new weapons in international warfare, etc. The controversial issues of planned parenthood, of eugenics, of euthanasia, of biological and psychological "engineering," should be discussed with an open and critical mind but also with a firm commitment to the value and dignity of each individual human being. Punishment will then no longer be regarded as retribution or retaliation, but wherever feasible it should be transformed into re-education, psychotherapy, and constructive social work. Then the basic moral commitments will become a matter of course instead of demands of a harsh and punitive conscience. Mankind will need to grow up to full adulthood; war will become an outdated means of settling conflicts of interest. Man's precious possession of reason and of good will has thus far been altogether only too insufficiently applied in the solution of his most urgent practical problems. We do have the powers of rational deliberation, of freely choosing among alternative avenues of action, and, philosophically and scientifically speaking, there is nothing in all this that demands either an indeterminism in behavior or a Cartesian action of a pure mind or soul on the body.

The quandaries of the traditional problems of mind-body and of free will vs. determinism have been among the most important obstacles in the mutual understanding between scientists and humanists. Once the confusions are avoided and the sting of the problems regarding man's place in nature eliminated, the road is open towards a genuinely scientific humanism. This I believe and hope will be a philosophy appropriate for our age of science.

NOTES

1. Hans Reichenbach, *Experience and Prediction* (Chicago, 1938).
2. For lucid recent accounts see C. G. Hempel, *Philosophy of Natural Science* (Englewood Cliffs, N. J., 1966) and R. Carnap, *Philosophical Foundations of Physics* (New York, 1966).
3. Cf. *British Journal for the Philosophy of Science* (recent years) various articles by Popper, Bar-Hillel, and others (on probability and induction).
4. Cf. especially P. K. Feyerabend, "Problems of Empiricism," in R. G. Colodny, ed., *Beyond the Edge of Certainty* (Englewood Cliffs, N. J., 1965).
5. Ernest Nagel, *The Structure of Science* (New York, 1961).
6. Cf. Ernest Nagel and James R. Newman, *Gödel's Proof* (New York, 1958).
7. Cf. Herbert Feigl, "The 'Mental' and the 'Physical,'" in vol. 2, *Minnesota Studies in the Philosophy of Science* (Minneapolis, 1958); J. J. C. Smart, *Philosophy and Scientific Realism* (New York, 1963); Wilfrid Sellars, "Empiricism and the Philosophy of Mind," in vol. 1, *Minnesota Studies in the Philosophy of Science* (Minneapolis, 1956); Paul E. Meehl, "The Compleat Autocerebroscopist," in Paul K. Feyerabend and Grover Maxwell, eds., *Mind, Matter, and Method* (Minneapolis, 1966).
8. Moritz Schlick, *Gesammelte Aufsaetze* (Vienna, 1938).
9. F. S. C. Northrop, *The Logic of the Sciences and the Humanities* (New York, 1947).
10. S. C. Pepper, "The Neural Identity Theory," in S. Hook, ed., *Dimensions of Mind* (New York, 1960); *Concept and Quality* (forthcoming).
11. Paul A. Schilpp, ed., *The Philosophy of Rudolph Carnap*, in Library of Living Philosophers (LaSalle, Ill., 1964).

12. Ludwig Wittgenstein, *Philosophical Investigations* (New York, 1953); Norman Malcolm, *Dreaming* (London, 1959).
13. F. S. C. Northrop, *The Meeting of East and West* (New York, 1946).
14. Max Born, *Natural Philosophy of Cause and Chance* (London, 1951).
15. Ernest Nagel, *The Structure of Science* (New York, 1961).
16. R. E. Hobart, "Freewill As Involving Determination, and Inconceivable Without It," in *Mind* (January 1934).
17. Herbert Feigl, "Validation and Vindication: An Analysis of the Nature and the Limits of Ethical Arguments," in W. Sellars and J. Hospers, eds., *Readings in Ethical Theory* (New York, 1952).

6

SCIENCE AND RELIGION: A REAPPRAISAL

John E. Smith

It has been some time since anyone has spoken seriously about the issue formerly known by the dramatic term "the warfare between science and religion" (or theology, as the case may be). The reason is that open hostilities of the sort that took place both in England and America in connection with the theory of evolution are no longer in evidence. To be sure, a sort of "cold war" has continued in some quarters, but there is today a better understanding than before of the sources of confusion and misunderstanding that plagued both sides, for example, in the classic struggles over evolution. More recently, the obvious impact of science on modern life plus renewed interest in the history and philosophy of science and the emergence of a new concern about the problem of *truth* in religion have had the effect of leading us to take a second look at the relations between religious affirmations and the deliverances of scientific knowledge.

It is not to be supposed that a new concern for the relations between science and religion necessarily means a return to the "warfare" of the past, though no doubt there are fundamentalists on both sides who hold either the view that scientific knowledge has made religion superfluous or that religion should ignore science because it is the handiwork of Satan. But whatever antagonism remains, we can say with some confidence that anyone who attended to the clashes between scientific philosophers and churchmen over the evolutionary theory of Darwin in the nineteenth century, for example, knows how completely rhetoric triumphed over both theology and science. As one wit remarked, the entire controversy might have been resolved if both sides had been willing to accept the view that "man is the great ape who made good!" Present concern is for a far more basic understanding, not only of the underlying causes that might have made such a dispute possible but also of the way in which the religious interpretation of reality is to be related to the scientific understanding of the world. Thoughtful men are always embarrassed at the thought of fundamental issues being resolved by default or, even worse, by indifference and neglect. However misguided were the disputants of the past century, they could not have been accused of apathy or indifference; they were prepared to fight for their convictions. If our failure to discuss the issues is a sign of indifference, then that is a mark against us and we are to be judged less candid than our forebears.

On the other hand, there is a new sense of the interconnectedness of things developing on the current scene. It is becoming increasingly clear that the religious interpretation of man and the world refers to the same reality about which our scientific knowledge purports to be true. If this is so, a theory of the relations between science and religion is unavoidable. Moreover, since religion is concerned with reality and is committed to truth in all its forms, men of faith will always endeavor to discover the implications of scientific knowledge for religion and to relate the content of faith to all the knowledge that can be attained. Men of science, it is true, do not always acknowledge a reciprocal obligation, but that is another question and it cannot be resolved here. There is, in fact, no "man of science" in the sense of someone who bases his total life, both intellectual and practical, on nothing other than the pronouncements of exact science. When the scientist as a total human person faces his own being and the attendant problems of his purpose and destiny as an existing individual, he becomes a man of faith—if not religious, then esthetic, moral, or even political faith. At the point where he makes the ground of his own conviction clear, he inherits the problem of relating scientific knowledge to whatever faith he has.

That modern science has transformed and to a large extent does determine modern life is a point too well known to be labored. This fact means that no aspect of modern life can remain out of all relation to science both as an enterprise or way of approaching the world and as a body of results. And, it might be well to indicate here that by scientific knowledge is not meant only physics, or indeed the physical sciences exclusively, but the whole range of empirical inquiry that can be conducted on an experimental basis and in a controlled way, including history, archaeology, and philology plus the spectrum of the so-called sciences of man. The ubiquity of science, both pure and applied, means that from any perspective and in defense of any interest you please, it cannot be ignored.

The first step in dealing with our problem is to make a distinction between those views on the one hand in which science and religion are brought into immediate relation with each other or are believed to confront each other "neatly," so to speak, and those views on the other hand according to which science and religion never confront each other immediately but only through the interpreting medium of philosophy (or theology as the systematic expression of religious doctrine). On the second view, distinct systems of ideas and aspects of experience can be compared and the relations between them expressed only through some generalized outlook or theory of reality as a whole. The distinction between these two views is itself a philosophical distinction, and it points to two very different ways of looking at the world and our experience in it. Every discussion of the relations between science and religion is at the same time a philosophical discussion and, in some cases, a theological discussion as well.

SCIENCE AND RELIGION AS IMMEDIATELY CONFRONTING EACH OTHER

Whatever philosophical position we hold, it is best to begin by acknowledging that we come to the discussion of our topic already in possession of a pretheoretical awareness of the fundamentally different intent behind science on the one hand and religion on the other. Science aims at attaining a theoretical, causal explanation of all things in terms that are essentially mathematical in character—although the degree of mathematical treatment possible will not be the same in all sciences—while religion sees everything from the standpoint of an ultimate ground and purpose for all things; religion seeks to diagnose the defect or flaw that separates existence from God and to offer man a resolution to the human predicament from the perspective of the divine. Religion is more concerned with *interpreting* the meaning and purpose of life than with *explaining* life in causal terms. The fundamental difference between the quest to explain everything and the quest for self-fulfillment or, in religious terms, salvation, is the difference that holds science apart from religion and marks out for each a domain over which it is master regardless of the other points at which the two are positively related.

In accordance with this difference of intention, it is necessary to admit at the outset that where a religious doctrine happens to incorporate or to be in some way dependent upon assertions about the world that are, from a scientific point of view, clearly erroneous, these assertions must be abandoned and the religious doctrine restated if necessary. In this sense the autonomy of scientific knowledge prevails and must be respected. On the other hand, the autonomy of religion must also be acknowledged so that, for example, the question whether and in what sense it is legitimate to speak of God as "personal" is one that is beyond the scope of scientific knowledge to resolve and, hence, cannot be decided except on the basis of religious and theological considerations. If, however, as I shall later argue, religion and science do not confront each other immediately, there will be very few cases in which we can say without qualification that a given religious assertion is flatly denied or contradicted by some proposition derivative from science. Such a situation would arise only where a literalistic conception of the Bible prevails and where it is thought that the Bible is a body of absolutely true propositions concerning every subject under the sun. But there is no need whatever to regard the Bible, which is essentially the record of religious meaning, in that way. The religious man will readily acknowledge that where religious doctrine depends on erroneous information or on what is demonstrably false from the scientific standpoint, an adjustment is necessary so that religion will consistently preserve its commitment to truth. Fortunately or unfortunately, depending on one's perspective, there are fewer cases of flat contradiction between religious belief and scientific knowledge than might be supposed. In the controversy over evolution, for example, the Darwinian theory explaining the origin of species was sometimes thought, by theologians and scientists alike, to contradict the Christian doctrine of creation taken in some wholesale and largely unspecified sense. But a more careful approach makes clear that contradiction is involved *only if* the religious doctrine demands for each and every identifiable species a separate act of creation understood in a sense that would preclude mutual influence in the lines of development. Even then, there would still be room for discussion, since we can never be sure that our concepts are absolutely precise and that the exact point at *issue* has been located and clearly expressed. The fact that Darwin himself regarded his theory as explicitly excluding separate acts of creation does not settle the question because the meaning of the creation doctrine has still to be made clear from the theological side. The point is that the religious doctrine is rooted in a tradition of thought that is constantly being reinterpreted; it is not to be identified with what some individual who has had no training in the matter happens to think it means or, even worse, what it ought to mean. The point of this line of discussion is clear: wherever we can confidently discover flat contradiction between a statement of religious doctrine and scientific knowledge, an acknowledgment and an adjustment must be made from the religious side. In any case, religion has no stake in falsehood and, in addition, the intention controlling all religious insight does not include theoretical inquiry into the constitution and causal determination of reality. Neither the Bible nor the creeds are handbooks of science.

To return now to the main point, there are two standpoints: one on the side of science and the other rooted in religion, from which science and religion are regarded as confronting each other immediately and with the result that each excludes the other entirely, since each is meant to cover the whole ground, so to speak, and the success of one is looked upon as in some sense the derogation of the other. The two standpoints in question have usually gone by the names of *positivism* and *mysticism* (or pietism), respectively. Although it might be supposed that no two outlooks could be further apart on the philosophical spectrum, the deeper truth is that they come together in the end, united by their common devotion to sheer fact, scientific and theological. The positivistic outlook bases itself on the fundamental proposition that the methods and conclusions of the natural sciences exhaust the field of knowledge and of truth and, consequently, that those aspects of experience that find expression in ethics, religion, esthetics, and indeed philosophy itself, fall beyond or beneath the sphere of knowledge. From this vantage point religion expresses no truth about the world except, perhaps, in the trivial sense that the human feelings expressed in or accompanying the recital of religious language might be verified as actually occurring. Whatever true propositions, if any, are contained in, or expressed in conjunction with, religious utterance are identical with statements that give information which, in turn, are identical with propositions in the special sciences. The religious perspective on this view is not rooted in objective fact but in feelings and emotions; religion may be elevating in its poetic grandeur or important as a means for aiding us in the performance of our moral duty, but it has no legititmate cognitive status. From the standpoint of the older and more strict positivist meaning criterion, the major portion of religious utterance is condemned as meaningless because, literally, devoid of sense.

I do not intend to consider the issue raised here nor to estimate the extent to which the positivistic outlook in its strictest form finds representatives on the current scene. More important is to understand the consequences for the relations between science and religion which this outlook entails. These consequences are three in number. First, religion and science have nothing essential to do with each other because there are no intelligible relations between the world of pure, objective fact disclosed by science and the world of feeling and fantasy in which religion moves. Second, the domain of the cognitive is coextensive with the pronouncements of the positive sciences, so-called, and all other forms of intellectual and cultural experience are reduced to feeling and convention with the final result that there can be no canons or criteria for judging religious statements in their own terms. Third, the problems in human life with which religion claims to deal are abandoned for solution to chance and to custom, or else finite and secular substitutes for our ultimate devotion are proposed in the form of morality, art, or politics. In the end religion cannot be in any sense rational, and science plus the world it discloses is separated from any transcendent meaning or purpose. There are two distinct worlds, one of pure fact and one of pure emotion, and they have nothing to do with each other.

These largely negative consequences stem from the assumption that to be significant or in any sense true, religious doctrines would have to be reached and supported in the same way as those in the natural sciences are reached and supported. The claim made on behalf of scientific knowledge to exhaust the sphere of the cognitive at once narrows that sphere and forces religion into emotionalism at the same time.

From the religious side we find that in some forms of mysticism, of pietism, and purely confessional theology the same disconnection of the two worlds, but this time it is the sphere of the divine that claims to cover the whole ground, and science is set at nought. Again the supposition is made that science and religion confront each other immediately and without either the need or the possibility of a dialectical or philosophical mediation between them. From these religious vantage points, science, and indeed all secular knowledge, is seen as irrelevant to, and in no intelligible way connected with, either the mystical intuition of the divine unity or the doctrine of God derivative from a revelation that is wholly discontinuous with man's cognitive faculties. From the mystical standpoint, scientific knowledge is finite and limited, belonging to the world that is evanescent and ultimately disconnected from the religious reality. No knowledge and no reality disclosed from the scientific standpoint has any bearing on the interpretation of the divine and indeed, from the mystical standpoint, no such interpretation is possible in literal, conceptual terms because that would result in finitizing God or reducing him to the status of a finite object subordinate to an encompassing natural order. The sphere of the mystical is that of insight expressed in symbols, not that of scientific fact expressed in propositions. Likewise, from the standpoint of present-day confessional theology, science and secular knowledge belong to the technological sphere, but they are not to be used in the elaboration of the revealed content of religion which must in any case be preserved from dissolution into secular perspectives. Nothing that is discovered about the secular world makes any essential difference to the religious truth so that, for example, historical and archaeological research into Christian origins, while such research is not suppressed, can occasion from the vantage point of the confessional theology no essential change in the content of theological doctrine.

We need not labor the point illustrated in the positions just indicated. Neither form of the positivist solution—and the religious standpoints mentioned represent positivism in theology—provides an adequate account of the relation between science and religion. Distinct the two may be, but distinguishing them is not the same as denying all relations between them. A solution that does no more than place the two in separate realms or that attempts to ignore one while absolutizing the other cannot be finally satisfactory. The religious problems remain and will do so as long as man is man; science remains and continues to progress and will do so as long as man remains. But the fact that man lives and moves and pursues his religious concern in the same world which science seeks to explain is the crucial fact that sets at nought all solutions that end with two non-connecting spheres. Science and religion find their own intrinsic natures in interrelating things; in addition they exist in the same world. It would be a most peculiar phenomenon if they should themselves remain unrelated except by the relation of total difference or otherness.

The chief defect of the two nonconnecting spheres solution is ultimately a philosophical one, a defect in fundamental conception. It is the error of supposing that two distinct and complex enterprises can be brought into immediate relation with each other. For to bring science and religion into immediate confrontation can succeed only in developing the features in which they differ most profoundly and through which their different intentions are made to stand in the sharpest possible opposition. If, however, we take a less abstract approach to the problem, we see at once that science and religion are related to each other, not immediately but in virtue of their mutual involvement in a whole of experience that transcends both. Science and religion each has its place in the structure of man's life in the world, and while neither is limited exclusively to man and his concerns, a theory of their relations becomes both possible and necessary because each concentrates on a definable aspect of the same reality. The positive connections between religious doctrines and exact, theoretical knowledge of man and the world are not immediately evident; they can be made clear only through a philosophical interpretation aimed, on the one hand, at discovering the possible bearing of new knowledge on ancient doctrines and, on the other, at discovering the implications of ancient doctrines for the proper use of such new knowledge. The office of philosophy is that of *interpreter* standing between religion and science, detecting relevances or irrelevances, locating genuine issues where they exist, and seeking at all times to bring each side to an understanding of the aims and methods of the other. The introduction of the interpreter is at the same time the introduction of the dialectical approach to the problem. We can now consider the relation between science and religion starting with the view that the two can be treated only in relation to a wider whole of experience philosophically interpreted.

SCIENCE AND RELIGION RELATED DIALECTICALLY

Science, in a fine description offered by Einstein some years ago, "is the endeavor to bring together by means of systematic thought the perceptible phenomena of this world into as thoroughgoing an association as possible."[1] Religion, on the other hand, is man's concern for an unqualified devotion to God understood as the ground and goal of existence as such; in contrast to science which can and must deal with finite and isolable systems, religion is concerned with the whole, the quality of human life as a whole, and of existence as a whole. The doctrines of religion and the conclusions of science have developed from different approaches to the world. Religious insight stems from an extended historical process of divine disclosure or revelation involving prophetic interpretation of the nature of human life and its ground. In the course of time religious insight finds further expression in the form of theological systems drawing on the resources of philosophy and secular knowledge generally. The conclusions of science, on the other hand, have been attained in accordance with that method of critical and cooperative testing which all associate with the empirical approach. In view of their obvious differences in intent and content, there is no ground for supposing that there *must* be a conflict between the two. In fact, apart from specific instances involving some particular theological doctrine and some particular conclusions of science that might be interpreted as involving a conflict, there is no need to make an *a priori* wholesale statement about a necessary conflict taking place. Since religion is committed to truth in all its forms, we can say that the religious view has no need whatever to obstruct scientific progress or to involve itself in the perpetuation of what is, from the scientific standpoint, false.

What most needs to be understood is that the supposition of a necessary conflict is a *philosophical* supposition, and in some cases it is a philosophical dogma. The conflict in question is not between science as such and religion but rather between a religious outlook on existence and a standpoint sometimes described as the "scientific outlook." This standpoint is not uniquely grounded in any single science or indeed in any one scientific law or theory, but is rather a generalized philosophical perspective purporting to express the way one should regard the world if one wants to be in accord with the scientific approach to reality. But, contrary to much that has been written and continues to be believed on this topic, scientific knowledge as such dictates the necessity of no single philosophical system. If the enterprise of science is generalized into a metaphysical position defining the nature of things as such, that metaphysics has to stand on its own in relation to alternative systems of the same logical type. The "scientific outlook" in the form of metaphysics does not enjoy a privileged position simply because it is supposed to be uniquely derived from science. Much mischief has been done by those who extend science as a method and a unified program into philosophy and then claim the authority of science for this philosophy which is not science. It is as if one constructed a mechanical horse programmed to run at least ten miles an hour faster than ordinary horses and then claimed that he was in fact an ordinary horse who just happened to win whenever he ran. If we can avoid the antagonisms and misunderstandings that come from the belief that science dictates some one philosophy—usually a naturalistic philosophy that is said to exclude God, final causes, and the whole range of lived experience—which obviously conflicts with the religious outlook, the way will then be cleared for a much more important and constructive enterprise, namely, the philosophical exploration of *specific instances* of interaction between religion and science for the purpose of becoming aware of at least three possible relationships that may obtain between the two.

There are three points at which philosophical mediation can serve to point up the relations between science and religion and thus create a community of understanding capable of containing possible conflicts and of turning them into viable discussion as over against a hostile silence. *First,* there is the possibility of pointing out the mutual contributions of each to the life of the other; *second,* there is the possibility of showing that advances in knowledge will make necessary the reinterpretation of old doctrine from the side of religion and also that having to interpret such advances from the religious standpoint will have the effect of drawing from religious insights hitherto undiscovered implications; *third,* there is the possibility that specific points at which conflict is alleged to occur can be analyzed more carefully for the purpose of locating an issue and formulating it precisely, so that we shall know whether in fact a conflict between science and religion occurs or whether we have instead a conflict between a philosophical extrapolation from science and a theological claim. The latter is not properly described as a conflict between religion and science; it is rather a philosophical dispute or a theological controversy.

The constructive approach here proposed relies heavily on one fact about the world and human life, namely, that process or development is real and that neither our knowledge nor the course of events is completely before us now. Religious doctrine is meant to apply to a world that is not now fully understood and is still in the process of being known; we cannot say in advance what future adjustments will be required in the interaction between religion and scientific knowledge. On the other hand, the full implications of neither religious doctrine nor of scientific knowledge are now known, and they can become known only through further interaction in which the attempt is made to see how ancient religious doctrine is related to novel information. The task that lies ahead is a creative and constructive one; its details cannot be anticipated or encompassed in a general formula. The task is best accomplished by treatment of specific instances rather than by wholesale pronouncements about what must or must not be the case.

The more precise meaning of the three relationships that have been proposed can best be given in the form of illustrations for each case. At the same time, by attending to the illustrations it will be possible to see what actually remains to be done in the way of analysis and interpretation. First, with regard to the contributions of religion and scientific knowledge to each other, two most striking and important instances can be cited. As a matter of historical record, one of the most important ideas that served to establish the scientific enterprise in the West was the idea of a universe that is regular and orderly in its behavior in virtue of its existence as the expression of a God whose will and power can be depended upon to be steadfast. In several remarkable, but strangely neglected, articles written some decades ago by Michael Foster of Oxford and published in *Mind,* one can find an account of the role played by this central religious doctrine in the establishing of the scientific approach to nature. It was no accident that the founders of modern science almost without exception looked upon their investigation of nature as an exploration or charting of the mind of God. We are accustomed to taking such expressions with a wry smile, at the same time congratulating ourselves at being free from the naiveté of great men who had not yet broken away from more primitive and quaint ways of thinking. But the fact remains that a basic religious idea was there at the beginning, and it guided those who had to voyage strange seas alone. Moreover, the idea of a truth that is objective and austere, the idea that for every specifiable question there is at least and at most one consistent and coherent answer, likewise finds its roots in the idea of God, who not only entertains the Ideas of Plato and the Divine Logos but who has the power to produce the reality in accordance with these forms. It is no accident that the great rationalist interpreters of modern science—Descartes, Leibniz, and Spinoza—all appealed to God as the foundation of the basic principle that, to use the formulation of Spinoza, "the order and connection of ideas is the same as the order and connection of things." This is not the place to enter into a dialectical discussion about the connection between scientific endeavor and belief in God's reality, but one point is clear: the supposition of the reality of certain grades of order and of the existence of an objective truth, both of which are essential to the continued pursuit of scientific knowledge, takes us beyond what any finite individual or any finite collection of them can directly verify in experience. Whatever the current state of the discussion, the historical contribution of religion to the scientific enterprise remains.

On the other side of the ledger, we must note that it would be difficult to overestimate the contribution of archaeology, of philology, and of various other forms of controlled historical inquiry to our understanding of sacred literature and the ancient monuments of the Western religious tradition. The so-called higher criticism of the Bible brought, indeed, its own problems of adjusting historical or theological meanings and interpretations, but no one acquainted with the record will want to deny that such study has given us new knowledge and insight into the meaning of the biblical record and especially has freed us from the unwarranted and needless supposition that the Bible was done at a stroke or that it "all fell down from heaven one night." The doublets, conflicting stories, anachronisms, and other signs of editorial work by more than one hand in the biblical books that have so delighted skeptics who have supposed that the whole religious outlook stands or falls with the accurate report of the death of King Uzziah, are all explicable through historical scholarship and a proper understanding of the circumstances conditioning the writing of the religious record. Here science has played its part in the freeing of the religious mind from superstitions and from needless suppositions that actually impeded our understanding of basic religious insights.

The point to be stressed is that insofar as a religious doctrine refers to a finite object, part of its meaning will be dependent on the nature of that object as disclosed from beyond the religious perspective. The incompleteness in our knowledge of finite objects implies an incompleteness in our understanding of the doctrine. When, for example, Augustine spoke of *man* and his sin, he could not have known that by "man" he should have meant what the modern thinker has to mean by "man," namely, the creature disclosed by the investigations of Darwin and Freud insofar as the results of these investigations are true. As our knowledge of the world progresses, religious doctrine purporting to refer to that world becomes related not to an entirely new reality, but to a reality with ever new aspects and features. With regard to the bearing of advances in knowledge on the reinterpretation of ancient doctrine, we may cite two instances relevant to this point. One is the discovery of the utter reality of time in the structure of all things as expressed in Whitehead's view that there is "no nature at an instant," and the other is the appearance of depth psychology as a vantage point from which to understand the behavior of man. Christianity has always taken the reality of history seriously and has never been in danger of collapsing the world into a timeless Absolute; despite its concern for history, however, Christianity has never fully made clear how God is to be related to the world of time and change in which genuine novelty and creativity exist. The emphasis on God as fully actual, as the One in "whom there is no shadow of turning" and who is always complete, has meant a reluctance to express in theological terms what has been called the "contingent" in God. The present state of scientific thought and the place of importance that is being given to time serve, and will continue to serve, as a challenge to religion to clarify its doctrine in this regard. The so-called necessary aspect of God will now have to be understood in such a way as to make possible some insight into the relatedness of the necessary to the emergence of novelty in experience and to the evanescent features of the world. Here it is the existence of scientific knowledge that forces the issue and makes it impossible for a theology that purports to interpret the world as it actually is to avoid taking time and change seriously.

On the other hand, in response to the challenge the religious thinker will return to his classical tradition in order to discover the resources it contains for understanding change and especially for estimating the limits of change and the degree of change that is compatible with identity, meaning, and purpose. In other words, while the new insistence on the reality of time and novelty sets a problem with regard to the interpretation of ancient doctrine, that same doctrine provides a vantage point from which an advance in knowledge is to be estimated and judged.

The second development—the appearance of depth psychology—poses a similar situation for theology. The new analysis of man and human behavior in terms of levels and complex psychological structures presents a picture of human nature to which the religious understanding of man must be related. On the one hand, the new psychology provides new and arresting insight into the ancient doctrine of man as a being alienated from God and himself and also as a being ever engaged in the misuse of his freedom leading ultimately to that inhumanity of man to man that has been one of the most tragic features of modern life. Moreover, the reinterpretation of freedom and indeed the possibility of its vanishing entirely at the hands of a psychology forced to maintain itself as a science and hence inclined at times to reduce human behavior to that of lower, less complex levels of life call for a response from the religious side. Christian thinkers are called upon to estimate the extent to which a deterministic explanation of human behavior is acceptable and at the same time to indicate in what sense human freedom is maintained within a religious tradition that stresses the absolute sovereignty of God in all realms of existence and predestination as an eternal decree laid down along with the foundations of the world. Other examples could be given that reproduce the same pattern of the emergence of new knowledge to which old doctrine must be related. St. Augustine, for example, surely one of the most profound interpreters of the human person who ever wrote, did not have to confront the Freudian, the Marxian, and the Darwinian man. Augustine's followers, standing in the same religious tradition in which he stood, must confront these modern interpretations; they must find ways to incorporate what is valid in these conceptions and at the same time interpret their significance in religious perspective.

With regard to the third function of philosophy in its role as mediator or interpreter between religion and science, the basic point has already been made and need not be labored. Not all the alleged "conflicts" between religious doctrines and objective knowledge of the world are actual conflicts. Before we can be sure that a genuine difference exists or an issue is at stake, we must be clear as to the meaning to be attached to the supposedly conflicting statements. The scientific statement together with its boundary conditions must be interpreted from one side and the religious doctrine must be stated and interpreted from the other. Only by following such a procedure will it be possible to know when adjustments become necessary and when the alleged conflict dissolves or is seen to be based on a mistaken assumption.

POSTSCRIPT

As a postscript, it might be well to call attention to a famous statement made by Einstein some years ago in a paper entitled "Science and Religion," delivered at the first meeting of the Conference on Science, Philosophy, and Religion in New York. Much that Einstein wrote in that paper has my approval, but he raised two difficulties about religion which recur so frequently that they should not be passed by in silence. On the one hand, Einstein expressed doubts about the doctrine that the Divine Will is an independent cause influencing the course of events with omnipotence, and, on the other, he called for the abandonment of what he clearly regarded as a closely related doctrine, namely, that of the personal God. Although he was not entirely clear on the point, it appears that Einstein would have substituted convictions about the reality of certain ethical values for the classical idea of God stemming from the Judeo-Christian tradition.

We cannot, of course, join the issues at length here, but two comments are in order. First, with regard to the omnipotent God who interferes with the order discovered by science, it has always been sound theological doctrine to hold that God affects the course of the world in accordance with the specific natures of the beings in it, including man who has reason and freedom. Consequently, the idea of God as an arbitrary Power capable of doing anything at all—the commonsensical conception of omnipotence (which, by the way, is a *real projection* on man's part)—is not the classical idea but the invention of religious fundamentalists who hold that God appears only in rational absurdities. With regard to the matter of the personal God, a more complex problem is involved. First, it would be necessary to explain carefully in what sense and why the concept of person has been regarded by the Judeo-Christian as the most adequate analogy or symbol for God, and, secondly, it would be necessary to ask why it is supposed that the doctrine of the personal God in a controlled formulation puts a greater strain on belief than some of the substitutes that have been proposed. One wonders whether it is more intelligible to claim existence for the Good, the True, and the Beautiful—as Einstein does—than for the living reality expressed through the *personae;* moreover, the idea of a cosmic order transcends sense experience as completely as the idea of a personal God. These are but suggestions; they are offered in order to call attention to the fact that such difficulties are not unknown to the spokesmen for religion, and they have often felt them as keenly, if not more so, than critics speaking either in the name of science or philosophy.

Whitehead has well said that the art of bare persistence is to be dead; neither religion nor science should be willing to solidify themselves wholly in dogma either of content or method merely in order to survive. One important way in which the two can remain living, however, is to continue to accept the possibility of fruitful interaction and mutual interrogation of each other. A science that is pursued with no awareness of its limitations soon becomes dogmatic and dies; a religion that tries to seal itself off, immune from the advances of knowledge and from critical inquiry, likewise dies. Human life can prosper as little with a dead science as with a moribund religion.

NOTES

1. "Science and Religion," in *Conference on Science, Philosophy and Religion,* vol. I (1941) 209.

7

SCIENCE AND INTERNATIONAL AFFAIRS

Ludwig F. Audrieth

It may come as a surprise to many that our government's concern with science and technology evidenced itself at the very beginning of our struggle to achieve recognition as an independent nation. As minister to France from 1785 to 1789, Thomas Jefferson reported regularly on European scientific developments to Yale, Harvard, the College of Philadelphia, and his Alma Mater, William and Mary. He was well aware of the scientific accomplishments of Lavoisier, the father of modern chemistry, and even went so far as to criticize as premature Lavoisier's attempts to set up a system of nomenclature for chemical substances. He argued the merits of chemistry with the great French natural philosopher Buffon (Georges Louis Leclerc) who had labeled chemistry as "cookery." Jefferson contended that "it [chemistry] is among the most useful of sciences, and big with future discoveries for the utility and safety of the human race." Jefferson was a very practical man. He collected seeds and plant specimens for shipment to the American Colonies. He was constantly on the lookout for new industrial and technological developments. He was one of this country's first practical natural philosophers, as demonstrated by his own flair for innovation and invention—and by the fact that he, as the first secretary of state, introduced and developed the United States patent system. Jefferson can be looked upon as the first science administrator and scientist-diplomat in American history.

Every school child associates the name of Benjamin Franklin with the discovery of electricity, but few realize the extent to which his accomplishments influenced international science in the mid-eighteenth century. Like all scientists, Franklin was anxious that others learn about his work. He therefore described his experiments in letters to English friends. Some of these letters were read before the Royal Society; others were published in the *Proceedings* of this learned organization. In 1751 he compiled the results of his work in the form of an eighty-page booklet entitled *Experiments and Observations on Electricity*. A year later it was translated into French. Revised editions appeared subsequently in England, with German and Italian translations in 1758 and 1774, respectively. Franklin became famous in Europe as a distinguished American scientist. His great reputation as a scientist most certainly gave prominence to his political opinions and pronouncements during and after the American Revolution.

These examples from early American history could be supplemented by many others dating back into antiquity. All such illustrations are witness to the fact that science has always been international in character and that scientific knowledge constitutes part of the cultural heritage of mankind. Even though we may lament the lack of communication between scientists and nonscientists, it is nevertheless an irrefutable fact that science, to quote Wiesner,[1] "is a true international language, one in which the ambiguities can be reduced to a universal dictionary available in all languages —nature—a code to be shared by all who will make an effort to learn . . . the language of science. . . . Its great advantage lies in the fact that the true laws of nature are the same everywhere, [and this is important] as are the problems of interest."

There is no limit to the problems which the physical world presents for study and research. These problems present themselves to all men wherever they may have been born and whatever their race or their national origins. The practitioners of science have always compared and exchanged ideas with one another. The accomplishments of great teachers have always extended beyond man-made boundaries. Great centers of learning have attracted students from all over the world in their quest for new knowledge. And this tendency to learn from others, to exchange ideas and experiences, motivated by an insatiable curiosity about the physical world, has made the scientific community a truly international community of the mind.

I

Yet there is another aspect of this search for "truth to gain an understanding of the world about us and ourselves." There is a second face to science, as Roger Revelle[2] has so aptly expressed it, which entails the "use of knowledge to gain control over nature and [what I consider pertinent to my discussion] power over men." It is for this reason that I would differentiate between science as a cultural pursuit and the applications of science through technology. It is science as manifested through technology that has advanced the state of civilization, that has helped man to adapt himself to his environment and has assisted him in meeting his material needs; but, what is equally significant, science has made it possible for the technologically advanced nations to achieve military power, economic strength, and a superior status in this highly competitive world. It is the scientific-technological revolution that has showered man with its benefits but concurrently has created serious problems which affect human relations at every level of organized society. It is science and technology that offer so much hope to all of mankind in contributing more adequately to the basic necessities of life, to higher living standards, and to affording more time for leisure activities but at the same time have given rise to political problems that extend beyond national boundaries and influence national and international affairs. Science and technology have therefore become matters of concern to the governments of all nations, since it is now recognized that scientific discovery and technological innovation impinge strongly on and affect markedly the development of both national and foreign policy and thereby influence the course of human events. Let me illustrate.

II

Even though technological innovations advance the state of civilization, it is probably less well recognized that such changes are also accompanied by political, social, and economic dislocations on the national and international scene. Let me document this statement by citing a few examples from history and suggesting problem areas which now confront us and which the future may present to us.

For thousands of years the dye indigo, so highly prized even by the ancient Egyptians, the Greeks, and the Romans, had been obtained from plants. Adolph von Baeyer succeeded in elucidating the composition of the natural product in 1883. Fifteen years later a commercial process was worked out to produce indigo synthetically. The result? Two years later synthetic production completely displaced the natural product. Cultivation of indigo and extraction of the dyestuff had become a major industry in India. Some 1,750,000 acres of land were under cultivation in 1897. India exported 1,700,000 pounds that same year. The economic and sociological upheaval following the sudden obsolescence of this source of livelihood affected an estimated 200,000 people in India.

More striking in its effect on the course of history was the development of a process for the production of synthetic ammonia. It was back in the 1840's that Liebig and others had shown that fertilizer materials containing the elements nitrogen, phosphorus, and potassium were needed to sustain and increase agricultural production. Particularly limited were the supplies of nitrogenous fertilizers. The problem? How to utilize the tremendous resources of nitrogen in the atmosphere for conversion into a plant food. Chemists the world over looked for synthetic routes entailing the fixation of atmospheric nitrogen. The electric arc process was developed. Methods for fixing nitrogen by calcium carbide were discovered. However, neither of these processes could compete on a world-wide basis with the huge deposits of saltpeter in Chile which in the meantime were being exploited for use as a fertilizer material and for the production of nitric acid.

Research demonstrated that nitrogen and hydrogen could be made to combine at moderately high temperatures and high pressures to form ammonia. But high pressure processes posed many engineering and technical difficulties. These were finally resolved by Haber, Bosch, and their coworkers. The first synthetic ammonia plant was placed into operation in Germany in 1910. Another discovery a few years earlier by Professor Ostwald demonstrated that nitric acid can be produced easily by the oxidation of ammonia. Why do I mention these facts?

First, nitric acid is an important raw material and is used in the manufacture of practically all conventional high explosives. Consequently, nitric acid manufacturing facilities constitute an important strategic industry in maintaining the military capability of a nation. The Allied World War I blockade had effectively denied Chilean saltpeter to the Central Powers, but the Germans now had sufficient nitric acid, produced by the oxidation of synthetic ammonia, to meet their military needs. The blockade did not reduce the potential and capability of the German Reich and its allies to wage war. One might hypothesize, after the fact, that World War I was prolonged by a deliberate German policy to encourage scientific research and technological development, admittedly in their national interest and, furthermore, dictated by their geographic position and lack of certain strategic natural resources.

Second, cheap fertilizer supplements were available for the more intensive utilization of land for food production. Attempts to starve out the Central Powers by imposition of the Allied blockade were at least partially circumvented.

Third, the technological know-how developed by the Germans became available to the Allied Powers after World War I. Other technologically advanced nations were then also in position to supplement their own chemical nitrogen requirements. The effect upon the Chilean monopoly promptly became evident. A scientific and technological development had broken a monopoly position and brought about serious internal dislocations in Chile which affected the people themselves and the financial solvency of the Chilean nation. The Chilean monopoly suffered a further blow in the early 1930's by the development of a process for making "chlorine without caustic." Nitric acid (from synthetic ammonia) and cheap salt were found to react to give chlorine and by-product sodium nitrate (Chilean saltpeter). Chile still has not recovered completely from this succession of technological advances.

But even today following World War II the Haber process still exerts a profound effect upon international relations. Millions of tons of synthetic ammonia are produced annually. Products such as ammonium nitrate and urea, both produced from ammonia, are shipped to the developing and highly populated areas of the world to enable these countries to meet more adequately their food requirements through increased agricultural productivity by the more intensive fertilization of the soil.

The story does not end at this point. The impact of a technological development may be delayed for years. Again, the Germans, because of their familiarity with the high pressure techniques, were able to develop a cheap process for making synthetic methyl alcohol, essentially from coal and water as raw materials. We had up until that time obtained our "wood alcohol" by a very crude process entailing the destructive distillation of wood. This impure, domestic product could not compete price-wise with the high-grade synthetic product. Again, a monopoly position was eliminated by science and technology, but not without its economic and sociological repercussions in two of our central states. Efforts to soften the blow by the imposition of a protective tariff proved futile to sustain an industry which could no longer stand on its own economic feet.

Many other examples might be cited. Let me mention just a few more that have to a greater or lesser extent caused economic and/or sociological dislocations. Synthetic fibres have affected the production and uses of cotton, jute, silk, flax, and hemp. Today over 75 per cent of our rubber needs are satisfied by various synthetic products. I would be guilty of an error of omission if I did not mention the outstanding accomplishments of one of Notre Dame's greatest scientists, the late Father Julius A. Nieuwland (1878–1936). His pioneering fundamental researches on the chemistry of acetylene served as the basis for the development of the first synthetic elastomer, neoprene, to be produced on a commercial scale in the United States.

One may wonder what the new synthetic top leather known as "Corfam" will do in time to beef and cattle producers the world over—who have always counted upon hide values as an additional economic incentive. What would happen if an acceptable synthetic coffee[3] were to become available on the world markets? What effect would such a development have on the peoples of Brazil, Columbia, and the Central Americas? What effect would the successful development of a process for utilizing algae as a food source have upon the economies of the large food-producing nations of the world? Some of us were startled by the announcement a few years ago that aerobic oxidation of petroleum in the presence of nutrient solutions containing only inorganic compounds leads to the formation of substantial quantities of proteins and amino acids. These products are now being evaluated as animal nutrients. The Finnish Nobel Laureate Professor Arturri Virtanen has demonstrated experimentally that cattle can subsist and thrive on a synthetic diet consisting of cellulosic matter and inorganic mineral nutrients. Is it unreasonable to speculate that synthetic human diets represent the next logical development?[4]

However, an even more important question arises from all these scientific efforts to augment our food supplies. Will these potential developments satisfy the subsistence needs of a rapidly expanding world population? New drugs and sanitary measures have reduced infant mortality and extended life expectancy. No one will deny that these efforts are worthy of commendation, even though the population problem is thereby aggravated. Is the answer to the overpopulation problem the development of birth control drugs? Can such drugs compete with centuries-old habits and religious traditions? Is euthanasia a solution? Or are we merely postponing the inevitable day when the population explosion reaches such proportions that food supplies will have become inadequate—and war, famine, pestilence, misery, and vice will restore the balance, as predicted by Malthus over 150 years ago? The food-population equilibrium has become a matter of vital concern to the advanced nations of the world.[5] Will synthetic foods solve the problem? Or are we faced with difficulties of a physiological and psychological nature as we approach a "standing room only" condition as the world population increases?

Every great scientific discovery and technological development exerts a world-wide influence. No monopoly position is certain. The price of progress is obsolescence. Science and technology recognize no status quo. Science and technology represent the very essence of change in upsetting traditional and entrenched patterns of life and living. It is the political, social, and economic adaptation to the changes wrought by advances in science and technology that represents the chief problem of modern society.

III

It can thus be demonstrated easily that scientific discovery and technological innovation usually bring about economic and sociological dislocations. However, the political effects are even more complex and diverse. And politics, divested of its sugar-coated pronouncements and high-sounding declarations to make it palatable for the mass of humanity, is essentially a matter of practical human relationships. It is the politician and the statesman who must advocate courses of action which contribute to the achievement of specific national and international goals. Marshall[6] defines foreign policy "as the courses of action undertaken in pursuit of national objectives beyond the span of jurisdiction of the United States. . . . Our foreign policy unfolds in the things done by the U. S. Government to influence forces and situations abroad."

Many of the central foreign policy issues which today affect our relations with other nations of the world have important, if not decisive, scientific components. Disarmament, space, nuclear weapons policy, and foreign aid are four of the more obvious examples.

Disarmament depends essentially on the assessment of the technological capabilities of potential aggressors. Since control through inspection procedures within the sovereign territories of various nations of the world is unacceptable (to the U.S.S.R.) or impossible (within China), we must resort to highly sophisticated technological systems to defend ourselves against a surprise attack or use technical monitoring devices to make sure that such agreements as the partial test ban treaty are adhered to.

Space exploration for the advancement of knowledge is one thing, but we should not forget that the nation that controls space controls the world. To what extent shall we make space technology available to our European allies for peaceful and utilitarian purposes? What are the advantages of cooperative programs with ELDO and ESRO? With the Russians? For scientific purposes and for the mutual benefit of participants in such programs we might well answer in the affirmative. But how far can we go in sharing with others our knowledge of guidance systems and booster technology when these areas are also vital to our national security—even though we cannot prevent such developments by foreign nations?

We early recognized that nuclear knowledge and the technological know-how associated with nuclear fusion and fission could be used to both our advantage and our disadvantage. We felt that it would be wise to restrict dissemination of information which might enable others to build up their nuclear weapons capabilities. Our policy was motivated by the desire to act as trustees of this knowledge for the world to prevent nuclear weapons proliferation. Yet at the same time we made it a matter of policy to advance the applications of nuclear knowledge for peaceful purposes. A new source of energy had become available, even though, at the time, not competitive with fossil fuels. The use of radioactive isotopes for medical and industrial applications has opened up tremendous possibilities for the good of mankind. We felt that these applications should certainly be encouraged in every possible way. But where is the dividing line between these policy objectives?

We have made nuclear fuels for power reactors available to other nations under bilateral agreements giving us inspection rights to prevent diversion of fissionable material for weapons uses. We have encouraged the establishment of Euratom (the European atomic energy group), of Britain and the outer six, of the EEC group nations in Europe that are pooling their resources for power reactor developments, and of the IAEA, the International Atomic Energy Agency in Vienna, to develop safeguards systems and to promote research and development on peaceful applications. But have these actions given us more than a breathing spell and a limited time advantage? After all, the basic science of the processes of fusion and fission is known to scientists the world over. We now realize that the monopoly position which we had originally attained has not been retained even with respect to our allies. We wonder who next is going to emerge suddenly with a test bomb. There is the concept of deterrence, the notion that such a thing is so horrible that no one would be willing to use it. But is this the only way that we can prevent the destruction of humanity, through deterrence?

Our technical systems program is motivated by both humanitarian and political considerations, but the interesting thing is that all the developing nations look to science and technology as means for achieving their national goals to give them security, economic strength, and a higher standard of living for their peoples. Here I want to quote Nehru in a statement to the Indian parliament back in 1958, the so-called Science Policy Resolution: "Science has given to the common man, in countries advanced in science, a standard of living and social and cultural amenities which were once confined to a very small privileged minority of the population." This was the basis of his request for the appropriation of funds in order to support a stronger scientific effort back in 1958. These technical systems programs require resources and capabilities which we and the other technologically advanced nations in the world must make available. To what extent should we divert our much needed scientific and engineering manpower resources to such assistance programs? Can we afford to send our teachers abroad to upgrade elementary, vocational, or university training in these newly emerging states when we realize we have a lot of homework to do; or doctors and nurses in greater numbers to battle disease and epidemics among the native populations; or agricultural specialists to increase the food production in the developing countries; or management experts to rebuild their governmental structures? Have we really made friends through the programs which we have already mounted? Or is all of this effort for naught anyway, as some authorities predict that the disparity between the technologically advanced and the developing nations will actually grow larger since enlightened self-interest demands that we and other advanced nations utilize our scientific resources and capabilities to maintain and retain our own positions of world leadership?

Foreign aid, nuclear policy, space, and disarmament represent political issues whose solutions politically demand scientific expertise. There is, however, another group of foreign policy issues affecting our relationships with the world at large that have emerged as the result of scientific and technological developments. These are less critical since they usually entail negotiation of bilateral or multilateral international agreements to set up administrative and procedural arrangements covering such matters as ground, sea and air transportation, allocation of radio frequencies to facilitate international communications, and the definition of the rights of coastal nations to the exploitation of potential sub-sea gas, oil, and mineral resources.

Another category of foreign policy problems is even farther removed from the causative scientific factors. Our Food for Peace program would never have been established had we not suffered from or been blessed by an overproduction of agricultural commodities brought about by the highly efficient use of fertilizers, herbicides, and pesticides and by increased mechanization on our farms to boost domestic production. Yet even such a program, undertaken for humanitarian reasons and offering great political opportunities, creates national and international problems. The Food for Peace program is directly tied in with our balance of payments problem. International credit arrangements are necessary or, if payment in soft currencies is involved, the utilization of such funds and the possibilities of third-country monetary or goods exchanges must be taken into consideration. Competition between United States and foreign shipping interests is accentuated. Policy questions are raised with respect to the sale of such surpluses to the communist countries, especially when some of our near neighbors and allies do not appear to worry about real or imaginary restrictions in doing business with the Iron Curtain countries.

Many of the changes brought about by science and technology have created problems and situations on the international scene which cannot be solved in a traditional fashion. A degree of understanding and integration of scientific and political knowledge is required today far beyond the relationship that was adequate in the past. The scientist has entered the political arena not only as a technical expert but also at the highest levels of government to assist in decision-making and in the formulation of national and international policies. What is more he is there to stay as a politically-minded scientist even though he may not qualify as a political scientist.

IV

Let me now discuss another aspect of science and international affairs which is more enduring and continuing in character. There is a magnificent idealism which motivates the scientist in his espousal of science as a "common enterprise" of man. Scientists have traditionally advanced the premise that science recognizes no boundaries with respect to race and language. Scientific progress is limited or contained by neither human nor geographical barriers. Scientists have traditionally shared their knowledge with others. These personal contacts the world over represent an opportunity as well as an obligation for all who call themselves scientists to go beyond their professional specialties in the hope that such contacts will ultimately build bridges of understanding and thereby help to relieve world tensions. And in this respect the scientist, like the writer, the composer, the artist, and the musician, can play an important international role. Anything that we can do through this "people-to-people" approach to develop a better understanding and tolerance for one another's points of view, for the maintenance of peace on earth, will be worth the effort.

All governments, including our own, encourage and support financially such international scientific organizations as ICSU and its member societies, UNESCO, WHO, WMO, FAO, and many other groups designed to facilitate cooperation on an international basis. Our government encourages the travel of scientists to international meetings, as well as the exchange of scholars and students through such programs as the Fulbright act and through various cultural exchange agreements. It encourages scientists to invite their overseas colleagues to the United States for conferences, seminars, colloquia, and meetings. All this is done not only to support science in its quest for knowledge but also because we believe that international science constitutes a forum to help promote international understanding. However, we should remember that scientific achievement has become a national status symbol with both advanced and developing countries striving to enhance their prestige thereby—and prestige has become a national goal and a political objective.

Many research efforts demand international collaboration. Such international programs are encouraged by our government. The Antarctica program was begun during the International Geophysical Year and continues to receive international support. Two major international projects are presently underway, the International Year of the Quiet Sun and the International Indian Ocean Expedition. The International Hydrological Decade will soon be underway with a very ambitious program. Astronomy, oceanography, meteorology and outer space exploration are fields of science in which the advancement of knowledge demands global cooperation and participation. The practical problems of health and nutrition, desalination and fresh water supplies, air, water and food pollution, and the development of engineering and materials standards come within the purview of the United Nations and such organizations as the OECD (Organization for Economic Cooperation and Development). The solutions to such problems will redound to the benefit of all mankind.

V

I have mentioned these areas in which international cooperation has been achieved to re-emphasize my earlier point that science is multi-faceted in its effect upon man and society. Scientific discovery and technological innovation do bring about changes and dislocations to which we must adapt ourselves. They do create serious political problems which plague our statesmen—whose political horizons are, however, limited to national horizons. Yet science also offers to us and to the whole world political opportunities for the maintenance of peace through the development of better understanding between peoples of diverse racial, religious, and ideological backgrounds. Through its applications it does offer the possibility of relief from hunger, want, and fear which are at the root of all the world's problems.

Science also represents part of the cultural heritage of mankind, and for that reason I cannot take seriously the "two cultures" polemic, a schism which exists only in the minds of frustrated, narrowminded specialists be they in the arts and the humanities or in the physical and engineering sciences. There is only one culture, which is to be likened to a forest of many trees—not a tree with many branches—each tree of knowledge growing and thriving as a biological entity but, nevertheless, dependent upon and affected by its environment and nurtured by the same soil of intellectual freedom which gives all disciplines in the Lord's Acres the conceptual and factual nutrient for continued growth. There is only one community of the mind whose unifying strength lies in its diversity.

Let me conclude my remarks by quoting from an address delivered by Adlai Stevenson[7] at the 1961 meeting of the International Astronomical Union.

> You [scientists] have given us dangerous powers, but we have not yet learned to control them. You have given us the tools to abolish poverty, but we have not yet mastered them. You may have given us the means to extend the span of life, but this may prove a curse, not a blessing, unless we can assure food, survival and then good health and a good life for the bodies and minds of our exploding populations. You may have made the world small and interdependent, but we have not built the institutions to manage it — nor cast off the old institutions which scientific progress has made obsolete.
>
> Every great change wrought by science is foreshadowed years ahead in the laboratory or on the drawing board. But it is not until the new device is fully built and functioning, and has astonished the whole world, that we begin to think of its human and political implications. We are forever running to catch up tomorrow with what you made necessary yesterday.

NOTES

1. Jerome B. Wiesner, *Where Science and Politics Meet* (New York, 1965) 23.

2. Roger Revelle in *Cultural Affairs and Foreign Relations,* Robert Blum, ed., the American Assembly, Columbia University (Englewood Cliff, N. J., 1963) 112–138.

3. The report entitled "Possible Non-military Scientific Developments and their Potential Impact on Foreign Policy Problems of the United States" constitutes an excellent source of information on this subject. It represents a study prepared in 1959 by the Stanford Research Institute for the Senate Committee on Foreign Relations.

4. See, for instance, A. R. McPherson, "Synthetic Food for Tomorrow's Billions," Bulletin of the Atomic Scientists (September 1965) 6.

5. The Second UN World Population Conference was held in Belgrade, Yugoslavia, August 30 to September 19 with 1,500 delegates in attendance.

6. Charles B. Marshall, Department of State Bulletin (March 17, 1952) 415–420.

7. See Department of State Bulletin (September 4, 1961) 402–407.

8

SCIENCE AND HUMAN WELFARE

Farrington Daniels

I

Science is a way of thinking and experimenting, and technology is its offspring. It is a challenge that attracts creative minds. The geographical frontiers are gone but, in the words of Vannevar Bush, "Science is the Endless Frontier."

There are many more scientists working than ever before. The statement still startles us that 90 per cent of all the scientists who ever lived since the beginning of mankind are living now. For example, there are a million chemists in the world today. The immense world-wide activities of these many scientists of the present are built on the recorded achievements of the past.

We have quick publication of current scientific researches all over the world and active cooperation between the scientists of every nation. Three million scientific papers are published each year. *Chemical Abstracts* alone cover over half a million research contributions in a year. Then, in addition, we have monographs covering rapidly developing areas of science which enable the investigator to undertake research with a head start in a new field. And we have textbooks that help to turn loose a stream of young scientists who each year are better equipped to make new advances. It is remarkable how effectively this system of reporting, abstracting, summarizing, and teaching goes on spontaneously in all fields of science—without any central planning.

In the United States and many countries of the world we have mass education and can build on a large base of students for the making of new scientists. Undoubtedly many potential scientists are lost after their high school education, but probably of those who go through college a large fraction who have outstanding ability in science are recognized and encouraged to take up graduate work and thus have an opportunity with financial support to undertake creative work.

We are fortunate in having two different kinds of scientists. Among them are the basic scientists who have curiosity as their impelling motive, who report their findings and do not bother to follow the long road toward useful applications of their discoveries. When a research is carried as far as they desire, they return to their laboratories and work on other pioneering problems in which they are interested, and again they feed their results into the world's pool of scientific knowledge. Such scientists predominate in academic laboratories.

Another type of scientist, found largely in industrial laboratories, is interested in science as applied in technology. These scientists try to adapt the results of pure science to human needs and to solve specific problems in order to improve products or increase efficiencies of manufacturing. The applied scientists often work in teams which contain scientists of different backgrounds, perhaps a mathematician, a physicist, a chemist, and an engineer. Rapid advances are made through this programmatic applied research, for example, the three short years from the discovery of nuclear fission to the first nuclear chain reaction, and a second three years from the nuclear chain reaction to the nuclear explosion.

There is no fixed line between basic and applied research: practical applications often require additional basic research in order to fill gaps in available knowledge, and many good ideas for basic investigations come from applied research. Applied research feeds on basic research, but basic research and its tools are also dependent on applied research. So technology grows.

Here I would like to make a personal remark regarding basic research and the wonderful life of a university professor in science. He has the freedom to follow his scientific pursuit wherever it leads, and he is kept alert and young by the stream of brilliant students who go through his laboratories and classrooms. I can think of no better job—except that of an emeritus professor who has all of the joy of a university but without its responsibilities.

I have come to believe also that programmatic research is an extraordinarily powerful institution which can solve almost any material problem if we put enough effort into it. In teamwork in research, both basic and applied, progress depends on the past experience of the scientists, and two minds with different backgrounds are more than twice as effective as one. The team approach in research has demonstrated its power, and it will be even more important in the future.

Another important factor in the phenomenal development of science has been and will continue to be the availability of new apparatus for making measurements. The perfection of electronic equipment has accelerated laboratory readings and greatly increased their accuracy; and mathematical computing machines are revolutionizing many branches of science, making possible the solution of heretofore insoluble problems and carrying out in minutes calculations that used to take months. The still greater use of computers will be an important factor in speeding up the development of both science and technology in the future.

We might try to pick out a few of the outstanding achievements in science during recent years before we try to consider what some of the achievements of the future may be.

Atomic energy with its radioactivity and isotopes has given us an extraordinary and unexpected new tool for use in the sciences. Its atomic bombs have profoundly affected the security of the world and the diplomacy of nations. Its production of electrical power is here already, and it will be a vital contribution to the civilizations of the future. In another field the development of genetics and the DNA code has been an astonishing event giving us an insight into evolution and life itself. Our advances in communications through radio, telephone, and television and our advances in airplane and other travel have shrunk distances so as to completely reorganize the relations between nations. Isolated islands of prosperity in a sea of poverty are no longer acceptable. In agriculture with fertilizers, pesticides, and applications of genetics, science has done such an effective job that in the United States we now have a huge surplus of food; but this is only a local and a temporary phenomenon. Medical advances have been spectacular with new knowledge and new surgical techniques, with cures through antibiotics, and the virtual elimination of former diseases such as malaria, yellow fever, typhoid, tuberculosis, and poliomyelitis.

The mere comfort of living has been developed to a high level. We push a button in any room and get the work equivalent of ten men for two cents an hour. Steinmetz, a half century ago, made the then fantastic prediction that we would some day simply turn a thermostat knob and get any room temperature we desired, with cooling in summer as well as heating in winter. Perhaps the future encouragement of bold risk research may turn up new, undreamed of applications, new tools, and new theories. Risk research, though vitally important, can be severely handicapped in industry if it is expected to bring in dollar returns to the stockholders on the research investment within a year or two; and risk research in academic institutions can be crippled if it must guarantee the graduate student a Ph.D. in three or four years or if it demands a published paper to bolster the immediate professional advancement of the senior researcher or to assure the renewal of a government research contract. Not every research project meets with success; if every one does, the investigator probably has not set sufficiently high goals for himself.

The nuclear fusion of the hydrogen of the oceans under controlled conditions to give useful electrical power (similar to that generated by the nuclear fission of uranium 235) is a goal being eagerly sought. I seem to sense somewhat less general optimism in this achievement than was prevalent a few years ago—but it may come. Exploration of space with and without human passengers is here—and it will be greatly intensified in the future. But we must not stray too far in our imaginations. When the time of a round trip to a distant planet exceeds the life span of the passenger, the passenger becomes superfluous. Control of the weather may come, in spite of complications and questionable successes of the past. One promising area here is in the restoring of some man-made deserts where partial restoration of the land cover can help to start downward circulation of air and tend to bring in moisture and a possible return of agriculture.

Certainly the expansion of electrical power from nuclear fission will increase. Direct conversion of chemical energy into electrical power in fuel cells will probably be important. The direct use of the sun's energy for heating, cooling, and power will come first in sunny, isolated areas. It will become increasingly important as our large reserves of fossil fuel become scarce.

We must face the demands for new materials as well as

for new sources of energy. *Pure water,* which most people have taken for granted, is running short. New sources of water could open up new land areas for settlement. Our search for pure water from salt water and from the air and from processed sewage will be greatly intensified.

Building materials have evolved from wood and stone to brick, concrete, and structural steel. New materials, in which plastics play a role, will be forthcoming. Our high-grade ores from which we get our metals and special chemicals are becoming depleted. Substitutes will be found, and particular efforts will go into the recovery from low-grade rocks and from the oceans. There is an inexhaustible supply of chemicals in granites and in the oceans, and these can be mined, although their concentration is measured in parts per million or less. Granites, when ground to powder, can theoretically yield, by chemical treatments, many different elements, including uranium, thorium, cobalt, nickel, tungsten, and copper. Both the water of the sea and the floors of the ocean will be called on to give up the chemicals which man needs.

As our population and our industrial manufacturing increase, pollution problems multiply—pollution of the land, of water, and of the air. These problems call for intensive scientific research starting now.

Research on aging will lead to a longer human life span. It is surprising that so little attention has been devoted to this field which concerns so many people. The theory of aging as being due to the cross-linking of proteins merits intensive research.

Will the progress of science slow down in the future? Or will it continue to expand at an ever increasing rate? At the turn of the century some unimaginative scientists were saying that all the laws of physics and chemistry were then known and that there was not much more to do in science except to improve apparatus and refine experimentation so that the measurements could go one more decimal point.

Then came X-rays, radioactivity, quantum theory, nuclear fission, and the expansions which we have just reviewed.

What can slow down the expansion of science? We can run out of men, money, materials, or incentives; or science may become so complex as to impede its own progress. For example, the literature may become so large that storage and retrieval of specific items from libraries will be too difficult. Will the development of science follow the well-known logarithm S curve observed in the growth of bacteria or the development of a biological colony? At first there are only a few units. They multiply, and the offspring increase almost exponentially and continue to multiply at an increasing rate until the food supply becomes inadequate or they produce some pollution products which poison their environment. The rate of increase at first is slow, then fast and faster, and then slow and slower. Platt has given a keen analysis of the situation and discusses the possibility of a slowing down of science particularly in the big expensive undertakings such as space exploration and high voltage atom smashers. He points out the splendid state of man in the future if the human race survives. However, the present rate of increase in the support of large science research projects cannot continue indefinitely. If it should, sheer extrapolation suggests that the expense of research and development could exceed the national income.

The continuation of scientific research at a high level depends on the support of science by the public. At present the support is generous and the enthusiasm high. The American way of life has always included an element of competition and exploration. The last few years have seen very effective advances in the explanation of science to the general public. But there may be a revolt against continuously expanding support for several reasons: competition of other activities for the national wealth, a reaction to the ever increasing complexities of a science-controlled civilization, or an objection to the emphasis on always bigger and deadlier weapons of war. Personally, I believe that much of the public realizes the importance of science and will continue to support it to the best of the nation's ability. But there are pressures from the ignorant and the anti-intellectual forces that could grow.

Will the incentives continue to be strong enough to attract new potential scientists? I think so in spite of the long training and rigidly disciplined studies which are necessary. The more impelling incentive is the love of exploration and the urge to push back the frontiers of knowledge. Then there is the satisfaction which a scientist gets in his own work—the thrill of seeing random experimental data fit onto a straight line or into a complex mathematical formula. There is an appealing beauty in scientific work and a satisfaction such as that which accompanies the creation of a work of art.

The financial incentive for scientists and inventors is not a matter of major concern. Many inventors love to invent whether or not they obtain financial rewards. Academic and industrial scientists get recognition and advancement through their work, but only rarely do they get specific financial returns for doing a specific research. It is true that the patent system of the United States has been a powerful factor in the high level of our industrial wealth and national income; but the incentive of patents is more important for bringing in private capital to assure the expensive development of an invention than it is to encourage the original inventor.

In such areas as AID and the Peace Corps a new incentive may be developing for some applied scientists and inventors who realize that raising the level of living in the poorer countries is an important factor in relieving world tensions. They realize that the economies of these rapidly developing countries, which are different from our own, may be affected in important ways.

On the other hand there is a "don't worry" philosophy

regarding the development of new applied science, which in my opinion might slow down its benefits. I encountered this philosophy in my early attempts to hasten the application of nuclear power and in my later attempts to hasten the application of direct solar devices. According to the "don't worry" philosophy, the technical ceiling will rise of itself, and present problems will be solved in the future. "Some magic material to fill the proposed needs will be forthcoming anyway, so don't spend too much effort on it now." But the technological ceiling does not rise automatically. It rises only because a few pioneers like to bump their heads against the ceiling and make it rise.

II

The abundance of material goods (such as food, clothing, houses, automobiles, factories, and offices) and services (such as transportation, communication, health, comfort, protection, and entertainment) are a measure of the wealth of a community. These are greatly affected by science and technology, with the result that there is a tendency for the rich nations to become richer and poor nations to become relatively poorer. When people are living at a subsistence level, most of their time and effort has to be spent on providing food and the bare necessities for living, and they do not have time to think of new ways of doing things. But once they develop beyond subsistence living, they can have machines to do the work of many men and thus create more wealth with which to buy bigger and better machines with which to produce still greater wealth. With this release from excessive routine drudgery comes the development of the arts and sciences and research, as well as the introduction of new technologies.

Individuals who create wealth share it with others, through trade or good will or income taxes. In the same way the rich nations share their wealth with the poor nations, and they are beginning to realize that the best way to assist is through helping to provide educational and technological opportunities so that these rapidly developing nations can create their own wealth. The influence of scientific research and technological advances can be seen not only among different nations but also among different industries in our country. Those industries, such as chemical manufacturing, electronics, and aviation, which rely heavily on research are booming; those industries which have not stressed research are not so prosperous.

The close connection between scientific research and wealth has been brought out prominently in recent years by the flow of research grants from the federal government. A considerable portion of these research funds has been granted to institutions on the east and west coasts of the United States. Around these laboratories have sprung up many new wealth-producing industries. Communities now realize the importance of these research funds as indicated in recent competition by eighty-six sites in forty-six states for the 340-million-dollar high-voltage accelerator which is soon to be built.

The wise distribution of government funds for scientific research is an ever increasing problem. Should most of the money be spent where the chance for productive returns on the research investment are most likely? Should the more effective institutions in research tend to get more of the funds and become still more effective? Should there be a balance between small and large institutions and between different geographical parts of the country? Should the less active research centers be encouraged to become more productive so as to spread the nation's resources?

The research struggle for new materials to meet special requirements goes on vigorously and continuously: steels for high-speed tools; lightweight materials which will withstand the fierce heat generated by re-entry of space vehicles into the earth's atmosphere; noncorrosive, long-lived metals and plastics with special properties, tailor-made to meet new demands. Always there is the urge for better, cheaper materials—and the demand is being met and will continue to be met.

New sources of energy are being constantly sought. In the past, industrial wealth has sprung up around coal mines, such as in England during the industrial revolution, in Pittsburgh, and around water power such as Niagara Falls. The picture has changed as transportation improves, and high voltage transmission lines carry electrical power for hundreds of miles, and pipe lines carry natural gas for thousands of miles.

We take abundant and cheap fuel and electricity for granted. The requirements for industrial power, transportation, communication, and comfort heating and cooling are now met chiefly with fossil fuels—coal, oil, and gas which were produced by sunlight geological ages ago. They will not last indefinitely. Atomic energy from the fission of uranium and thorium will last much longer, but uranium and thorium, like coal and oil, are limited. The sun's energy is continuing and ample. The amount falling on a large desert in northern Chile in a year is equal in heating value to all the coal, oil, and gas burned up in the whole world in a year. Solar energy is sure to work in areas where sunshine is abundant, but it is now too expensive to collect and use in competition with cheap fossil fuel. But there are places in the world where fossil fuel is expensive and where the sun could be put to work now. Almost no support of solar energy research is presently coming from the federal government. For the past decade the Rockefeller Foundation has supported a considerable program of solar energy research at the University of Wisconsin, a program stressing help to the rapidly developing nations. We are grateful for this support. A recent book summarizes the present status of different ways of using the sun's energy directly, points out research areas, and gives bibliographies. In the USSR about five times as much effort is going into solar energy research as in the United States. In view of its importance for the future, solar energy possibilities should be studied much more vigorously.

For small units of a few gallons to a few thousand gallons, solar distillation is the cheapest means for getting fresh water from salt water. Over two million gallons of fresh water per year are now being produced by solar distillation. Although only a few solar devices have been successfully introduced as yet, it is expected that with new materials and new research developments there will be, in the future, considerable activity for heating, cooling, and the generation of electricity through heat engines and thermoelectric and photovoltaic generators. The very expensive solar generation of electricity by solar batteries has suddenly and unexpectedly become very important for use on space vehicles. It will become increasingly important.

Other areas of solar energy research include improvements in the design of exterior surfaces for homes and offices to control the heating or cooling load, as well as the development of materials which will withstand long weathering and exposure to sunlight. Exotic types of agriculture will be developed in which food crops will be grown at high efficiencies. Uncertain beginnings have been made with the mass culture of algae. It is important also to grow plants in desert or semi-desert areas under plastic covers in such a way as to conserve expensive water distilled from the ocean or from brackish wells. Serious problems such as overheating are involved. Long-range research should be started now for meeting future food shortages with new nonconventional types of agriculture which have potential for doubling or tripling yields in special areas rather than for increasing them by mere tens of per cents. Already it is possible to make edible foods from petroleum and coal. These researches should be pushed. It is interesting to realize that even now in our conventional machine-operated agriculture for every calorie of food produced another calorie of petroleum fuel has been used for ploughing, harvesting, processing, and distribution. When we run out of fuel in the distant future, we will run out of food also—unless the scientists make suitable preparations.

Traditionally, scientists have been happy working alone in their isolated laboratories. It was no concern to them if their experiments led to the obsolescence of an industry or to a new weapon of war. But all this was rudely changed when the scientists developed the atomic bomb. I was in a position of administrative responsibility then and found myself as a buffer between the army which emphasized the importance of complete secrecy and the scientists who were aroused to the political, social, and moral implications of the bomb. In July, 1945, at the request of Dr. Arthur Compton, I took a poll of the scientists at the Metallurgical Laboratory of the Manhattan Project in Chicago as to how the atomic bomb should be used in the war with Japan. The scientists eagerly discussed these implications among themselves and appointed committees. When the bomb was exploded and the secrecy ban was lifted at the close of the war, they sought to enlighten the general public. They flocked to Washington at their own expense and on their own time to educate congressmen and to lobby for the civilian control of atomic energy. They founded the *Bulletin of the Atomic Scientists*.

The sense of social responsibility is seen in many scientists. They are greatly concerned with the possibilities of war, as well as with the welfare of their neighbors, the economic improvement of the poorer nations, and an adequate supply of food, fuel, and resources for the future.

Many persons in all fields of thought feel that the uncontrolled increase of the world's population is ominous. At last its seriousness is being widely appreciated even in our prosperous country where new schools are needed daily and university dormitories become skyscrapers. The need may be much more obvious in the overcrowded poorer countries whose leaders are learning that something can be done about the population problem.

What advantage is there if the gains in agricultural productivity are immediately wiped out by explosive increases in the number of mouths to feed? There are three billion people in the world now. By the year 2000 there will be six billion, and not even standing room will be left eventually on the planet if the present exponential rate of growth continues indefinitely.

The problem is a complicated one, but science has done an effective job in providing means, medical or mechanical, for controlling the population increase. Wise parents can now limit the size of their families so that their children can have the best opportunity for development and education and the general community can be spared the burdens which parents of unwanted or too numerous children will not or cannot assume. It is now up to the governments around the world, to the leaders of thought, and the institutions of culture to encourage the general acceptance of family planning.

When one's nearest neighbors are far away, people can be careless with their disposal of used materials; but when frontier freedom is followed by urban crowding, all the efforts of the scientists, engineers, and governments are required to maintain a healthy, productive, and pleasant environment. Fortunately, here too the people are becoming acutely aware of the need for improved methods of waste disposal—again science can greatly improve conditions if given a chance. Sewers, garbage trucks, and beautification programs are improving our land. Sewage treatment and control of water contamination from manufacturing operations are trying to save our rivers and lakes. Now we are having acute problems of air pollution, as often mentioned—dramatically with smog in Los Angeles. Scientists have shown that complicated reactions are involved between the nitric oxide of the exhaust from millions of automobiles and organic matter in the air, accelerated by sunlight. Scientists will continue to explain new pollution problems of the future. They will try to keep ahead of the troubles produced by detergents and pesticides and new yet-to-be thought-of chemicals used by a civilization with ever increasing demands. But the governments and the people will have to follow the rules set up by the research scientists and the engineers. Fortunately, many of the pollutants are organic in nature and can be rendered harmless, in time, by the oxygen of the air and sunlight. Radioactive wastes do not respond to these natural chemical methods of purification, but here again scientists are at work and will be prepared to provide safe procedures.

Our present international tensions have been largely between East and West—between capitalistic and communistic nations. These tensions seem to be easing (except for communist China). The differences become less as the private-enterprise nations distribute the wealth through income taxes and social legislation and the USSR finds that incentives for private initiative are helpful. The future international tensions may be more along the North-South axis —between the richer nations and the poorer nations, most of which are located within 30° north and 30° south latitude. This area happens also to be the "sun belt" where solar radiation is an important natural resource. The people of the nations with low economic income can see that a better life is possible for them, and they are demanding it *now*. It is vital that the rich nations help the poor nations. Over the years the church missionaries have been devoted and helpful, and now the Peace Corps is proving to be effective. Scientists and engineers are beginning to point the way to new technologies suitable for local conditions and economic systems which will raise the standard of living. These efforts in the future will be sorely needed to reduce the economic gap and the resulting political tensions. We have seen the remarkable achievements in science and the extraordinary influence that they have had in developing our present civilization. Some of the methods which have done so much to advance science are being used by social scientists and those who administer human affairs. Our electronic, mathematical machines will be a valuable aid since human problems are of such great complexity. We have commented on the industrial revolution which followed the introduction of machines and expanded the power of human muscles; we may be facing a still greater revolution now that the introduction of electronic computers expands the power of the human mind.

Science has the power, from a *material* standpoint, to solve many of the problems of the human race. It can help to provide for most of the peoples of the earth an affluent society which can be even much more advanced than the society which we know here and now. But it can also create serious problems. The solving of human problems lags far behind the solving of material problems. With science we can surely solve our material problems; with *time* and good will, we can solve our human problems.

Science can help to solve our future difficulties, but the important solutions lie with the attitudes of the people themselves. Attention was given at this Symposium to the relations of science and religion. Here let me say only that unselfishness taught by religion over the past 2,000 years has profoundly affected our human institutions. We have no control to show us what our civilization might be today if we had not had this influence of religion. We may look eagerly in the future to reports of civilizations on other distant planets where the conditions for living creatures are similar to those existing on our earth. With different religious and moral backgrounds how have such peoples (if there are any) evolved and solved their problems of living together? The creeds and the rituals do not seem important to the socially minded scientists, but the influence of religion in unselfish personal relations and service and in the organization of human institutions is tremendously important.

There is one problem which, in my opinion, transcends all others—that of atomic warfare. The *major* effort of the scientists, the policy makers, and the general public should be directed toward reducing the chances of war. In atomic war there is no time, and the human race has the physical power to commit suicide and turn over the inhabitation of a radioactive earth to the cockroaches and the insects which can live in a world of higher radioactivity than could be tolerated by human beings.

One of the most vivid recollections of my life turns back to the day in August, 1945, when the newspapers announced the atomic bombing of Hiroshima. On a street corner near the Atomic Laboratory in Chicago I said to another scientist-administrator, "This is a sad day when we have put dynamite into the hands of children." He replied, "What do you mean, sad day? Every man, woman, and child in this country has been doing his utmost to end the war. Now you scientists have done it." "Yes, we've saved our sons, but I'm not sure about our grandsons."

As we look to the next hundred years, let us realize that our civilization will continue to depend more and more on science and that it will be abundant and happy *if* our political officers and public leaders, our Christian and religious spokesmen, our scientists, and the people of all nations join intelligently and vigorously in applying present science and new science to help all people and to operate human institutions—international, national, and local—in a framework of law and justice. The "if" is a *big* "if."

The betterment of mankind, and in fact the *survival* of mankind in the future, depends not on science and the scientists but on the people who apply science to human affairs. In the past hundred years intellectual and economic man has produced an unbelievable expansion in his understanding and exploitation of his environment. In the next hundred years moral man may well produce an unbelievable expansion in his understanding and development of himself. There is now a law operating in human affairs which needs far greater attention than it is presently being given—the law of human cooperation.

BIBLIOGRAPHICAL NOTES

There is a wealth of fascinating material on the subject of science and the future welfare of mankind in recent books and magazines.

There are three significant publications by the National Academy of Sciences.

The Scientific Endeavor (New York: Rockefeller Institute Press, 1965). This book celebrates the founding of the Academy a century ago. The last chapter is a notable address on a century of scientific conquest by President John F. Kennedy—one of his last public addresses.

The Academy's report to Congress on *Basic Research and National Goals* (1964) under the leadership of G. B. Kistiakowsky includes the different viewpoints of many leading scientists.

Another Academy report on the *Growth of World Population* is available.

L. V. Berkner's *The Scientific Age: The Impact of Science on Society* (New Haven, Conn.: Yale University Press, 1964) discusses many of the topics of this Symposium.

Harrison Brown's *The Challenge of Man's Future* (New York: Viking Press, 1954) is indeed a challenging book with emphasis on the possibilities of the application of science to human problems, prominent among which is that of excessive population.

R. E. Lapp discusses in his recent book *The New Priesthood* (New York: Harper and Row, 1965) the importance of science and the responsibility of the scientists.

D. Gabor in *Inventing the Future* (New York: Alfred A. Knopf, 1964) points out that civilization faces three dangers—destruction by nuclear war, overpopulation, and an age of leisure.

J. B. Wiesner's book *Where Science and Politics Meet* (New York: McGraw Hill, 1965) discusses science opportunities and problems with his authority as science advisor to President Kennedy.

H. Jarrett, ed., *Science and Resources: Prospects and Implications of Technological Advance* (Baltimore, Md.: Johns

Hopkins Press, 1959), published for the Resources for Future, Inc.

M. Polanyi in *Personal Knowledge* (Chicago: University of Chicago Press, 1958) stresses the thesis that the scientist's personal participation in his knowledge is an indispensible part of science itself.

C. P. Snow in *The Two Cultures and The Scientific Revolution* (New York: Cambridge University Press, 1959) stresses the difficulties in and the importance of the communication and understanding between scientists and nonscientists.

E. Rabinowitch has written a significant book on *The Dawn of a New Age* (Chicago: University of Chicago Press, 1963).

R. B. Lindsay, *Role of Science in Civilization* (New York: Harper and Row, 1963).

H. Boyko, *Science and the Future of Mankind* (Bloomington: Indiana University Press, 1961).

L. M. Marsak, *The Rise of Science in Relation to Society* (New York: MacMillan Co., 1964).

J. Ellul, *The Technological Society* (New York: A. Knopf, 1964).

C. P. Haskins, *The Scientific Revolution and World Politics* (New York: Harper and Row, 1964).

H. C. Benjamin, *Science, Technology and Human Values* (Columbia, Mo.: University of Missouri Press, 1965).

R. Schrader, *Science and Policy* (New York: Pergamon Press, 1963).

R. Gruber, *Science and the New Nations* (New York: Basic Books, Inc., 1961).

G. Piel, *Science in the Cause of Man* (New York: Alfred Knopf, 1961).

The American Academy of Arts and Science has published in its Journal *Daedalus* two symposia, "Science and Technology in Contemporary Society" 91, 2 (1962) and "Science and Culture" 94, 1 (1965).

The United Nations has a regular publication on *Impact of Science on Society*.

Seminars on Science, Technology, and Foreign Affairs have been published by the Foreign Service Institute of the United States Department of State by R. A. Rettig (1964) and L. F. Audrieth and H. I. Chinn (1965).

The *Bulletin of the Atomic Scientists* under the wise and effective leadership of Eugene Rabinowitch has had many challenging and authoritative articles on science and public affairs. Its chief theme has been the absolute necessity for preventing a nuclear war.

The issue of November, 1965, had a pessimistic article by Max Born who expresses the view that the scientific method cannot be properly absorbed by our civilization. We are not ready yet for such a powerful tool. He sees political and military horrors and a breakdown of ethics as a necessary consequence of the rapid rise of science. "If we survive a nuclear war," he says, "the human race may degenerate into a flock of stupid, dumb creatures under the tyranny of dictators who rule them with the help of machines and electronic computers."

These are only a few of the many interesting published contributions to the field.

They vary greatly in their range of optimism and pessimism. I count myself among the optimists. The pessimists stress chiefly an atomic explosion and a population explosion. In my opinion, the greatest danger lies in the possibility of atomic warfare.

9

SCIENCE, EDUCATION, AND THE FUTURE OF MANKIND

Philip Morrison

First of all, I would like to make some substantive prophecies. I do not make these prophecies in the sense that they are likely to come true. I know very well that the domain of the possible is much greater than our imagination, at least my imagination. I should like, however, to point out that I eagerly embrace the opportunity to present this prophecy in the sense of tentative groping. A prophecy—which is a projection—is a better calibration, a better insight, into the unstated premises and concerns of the speaker than almost anything else that he can say. Therefore, instead of trying to explain my experiences and how I have come to feel, I will make some bold prophecies. Once you see the trend of these compared with your own view of the future, you will see what kind of premises, what kind of preconceptions, what kind of errors and misapprehensions I work from. Were I to do otherwise there would be no way to judge the validity of what I say, and since there is not enough space to present the indispensable evidence, this is the best I can do.

First, I want to make a very simple, direct statement, one I really do believe will come true. That doesn't make it more likely to come true, you understand, but it shows I have a feeling of confidence. (Again a statement about me, which readers may take into account in their necessary evaluation.) This Symposium took place in a fine new hall after a hundred years of science at Notre Dame. Now, I venture to say that there are certain important differences between this hall and the hall that might have been used a hundred years ago to speak of a century of science in prospect at Notre Dame, had such a speech been undertaken. But the differences are small. They are differences a little bit like what is in the American bazaar of consumer goods where enormous novelties are promised us, though often they are not really very novel. When they are novel, they are novel more because they save money for the manufacturer than because they bring a new pleasure or power to the consumer. Direct dialing is sold as a great new advance; it is very important to the telephone company, I have no doubt, but for me it is a distinct if small step backwards, and everyone who has used the telephone knows that.

I would like to ask the question: What is in this hall that might differ in a hundred years? Here I think we are going to come to a true watershed. The room is more brightly lighted than it could have been in 1865. The heating system is different, although it was heated then, I am sure; perhaps it was a little draftier with a coal stove, but that was primarily a producer's concern. The man who kept it hot had a different task than the man who keeps it hot today; he is keeping more rooms hot per man hour, but that is not our problem—that is the problem of the auditors. As the "consumer" sees it, as hearer or speaker sees it, change doesn't come out very well. There were candles and there were oil lamps then, I suppose: excellent sources of light, nearly white, clean enough, nothing wrong with them. Not as cheap, not as controllable as today, but we don't use any of those controls; no one is adjusting the lights, no one is spinning colored wheels, and none of the opportunities which electric light has made available to the communication of idea and mood is used at all in our context. Our context, if you will pardon the colloquialism, is desperately "square," that is to say the educational context, the context of serious persons addressing themselves somewhat formally to intellectual issues. It has not benefited much from that novelty, not always good, but characteristic of the arts in the twentieth century. I would even say that novelty is characteristic of the science of the twentieth century, but it is not yet characteristic of discourse—university, academic discourse. This I think will not persist.

A hundred years from now an auditorium will be a much richer one. I see this in the following sense. I didn't even bring slides; had I brought slides I could have managed to get them displayed, I think with some pain, on a screen that would have appeared behind me. But I didn't because I know it is hard, and they weren't sharply relevant. But, of course, I can refer to books, to pictures, to devices, to children, to many things. If I could at the touch of a wand produce the images, the pages of books, the voices of speakers, I could make a much more convincing and certainly a much richer display. Granted, I might lose the austerity of the commitment of a single speaker (you always gain something for what you give up, you always give up something for what you gain) but I could strike a much better balance. We don't do that today. A debased form of intellectual communication, the television screen, does this at great expense. Why? Because the mechanism is still very complicated. For me to do that would require stage management, directors, producers, lots of work. That would not be true in a hundred years.

I venture to say that toward the end of that century an auditorium as well appointed in a university as central as Notre Dame will have a black box in the back of the room, holding access to the entire recorded body of knowledge, information, and image that the world contains. One will be able to summon it up at small expense by punching some keys; any page from any book will be shown projected immediately. It will surely be as easy as that, or easier.

I feel that science and its application to technology has not yet begun to change the life of the mind anywhere near the degree to which it will one day rise. What the printing press did in the beginning was not only to spread the growth of knowledge from a few clerks to a large body of persons, but it made possible as well the division of intellectual labor to a great extent. In the future for all intellectual activity, the machine, the computer, must be included. A new person will enter the world. A hundred years from now I would have a much richer way of speaking. Having punched a few keys, produced a little card for myself, I could simply press the card down on the table on a few contact points. On the screen would be displayed very nicely in color, at will, any one of ten million books, any section of any motion picture, or any previous summary which I might like to show. There is no reason at all why we should not have this sort of thing. It will come. The technology is on hand.

If I worked for the Central Intelligence Agency, I would already have nearly such a thing for these thirty million mimeographed documents that they have. This reference, deliberately ironic, is to show that I don't mean that what will come up on that screen will always be true or sensible, not at all. That is a problem for human choice. Just as all that is printed does not do justice to the memory of Gutenberg, so all that which will be displayed by computers will not do justice to their power. I am methodically of the opinion that we are in a state of transition; we are going to see changes—we here will not see them but we can begin to feel them—greater than anything else in the history of mankind for eight or ten thousand years.

Let me make another remark just to show the kind of

science that I think will be characteristic of the century (it is not my job to tell about this, but I trespass just to calibrate myself). Biology will change notably. The most important change in biology will be, I think, that there will be four, or perhaps five, independent kinds of life. One, our own life, that continuity with the tobacco-mosaic virus which we share even to the handedness of our amino acids. Two, a kind of life which will be evolved out of our life, and therefore not disjunct from it, but distinct from it, with zero chemical or hereditary affiliation with any other form of life. This will be what I would like to call the life of the machine, namely, electronic, crystalline, etc., which we will manufacture in some way. Third, a kind of life which is chemically recognizable as similar to our own, but which we will have made in the test tube by some special means, not very advanced, but quite interesting. So, there will be three kinds of terrestrial life. Fourth, I think there will be exobiology, the biology of independent, naturally appearing life spheres, most probably on the planet Mars; fifth, with a small probability, but I do not exclude it, some remote planet may make itself known by a communication revealing large-scale intelligence. This is the picture which biology will face a hundred years from now.

I would say most soberly that I do not think I have visualized what *will* happen, but the general picture that I should have evoked, the metaphor my guess contains, I think will be real. We cannot produce the precise future but we can say, broadly speaking, what sort of future we are going to see. I think by this path to escape the logical contradiction threatened by Professor Feigl, who points out we can't predict future discoveries without making them present ones. That is true, but I can foresee theoretical ones which might be discovered and regard them as true. That is nothing of a contradiction; the predictions are not real, but nevertheless they convey something. Approximation is the principle of the physicist.

What else will be new in the world? I think there will

be two or three other things which are relevant to the discussion of education. First, I think that it is inexorably the case that other national groups than our own (or the ones of Western European tradition and of West Asia) and very many more individual persons, very many more minds of all kinds, will be brought into the domain of science than has ever been envisioned. I think that contributions fifty or seventy-five years ago, limited to a few languages of Western Europe, will find themselves spelled out in the major languages of the world; the kinds of persons engaged in this, a very small set in Western Europe 300 or 400 years ago, will be found widespread over the entire globe in many societies, in many countries. I suspect that the academic world—and especially that part of the academic world which like ourselves addresses itself to abstract questions, having very little to do with the tasks of the professions or even with specific technological advances—will be a very much larger fraction of mankind than it has been before. As we today see learning to have grown from printing, we will see a rich new growth of learned works. It is a frightening prospect perhaps, but I don't believe it can be avoided. I am not trying to make a secure prophecy but only to show what the continuity of tendency suggests.

In short, I feel that we stand at a great time of change, a time of change less deep than only one previous time in prehistory when people changed from the way of hunting to the way of farming as the essential means of life. I suggest that pastoral and agricultural images are still the ones that control our minds. We still stand in the relationship of sheep to shepherd. We partake of bread, we prepare tables, we are concerned with cups which may be filled. These are artifacts acting as symbols which we still hold, which were not present ten thousand years ago. The raw material of such metaphor did not even exist for the men of ten thousand years ago. They did not know much about cups, about sheep, about shepherds. This whole change came, as we know now, most probably 10,000 to 8,000 years ago in Western Asia, in the foothills surrounding the great valleys of Mesopotamia where the earliest civilization grew 5,000 years later. I submit that the changes we are now undertaking are equally as great and, I think, faster than those of that time. Because we are immersed in them, we are not prepared to see their magnitude.

Let me again put it so. When we teach school children, the image we use is the image of the farm and the farmer, the crops and Farmer Jones, and the red cow. We recognize this is in fact the substantial basis of our life. But any glance at economic statistics will demonstrate that the way of life which is called farming is becoming rarer and rarer in this country, and extrapolation will show that in twenty years, such a short time as that, the number of persons who work in laboratories will be greater than the number of persons who work on farms. That is almost at hand, yet it is a change so extreme that compared with the history of man, it deserves enormous attention. It lies in the background, not merely in an economic way, not merely in the way of income taxes, but in the way of metaphor, language, symbol, terms of thought, values. The world is going to be immersed in, consumed by—if you wish to be antagonistic—science and technology; we must come to grips with that. Our own time is in this great transition.

I would like to assign a date to this great transition. The growth of modern science dates from the thirteenth and fourteenth centuries, from the tradition of theological skepticism, which culminates in some way with Nicholas Cusano's *On Learned Ignorance*. Here might very well be traced the beginnings of our modern science. From 1400 it will be 600 years by the time this century is over; and 600 years represents an enormous change, from talking on a purely philosophical level about how problems should be approached for which no certain answers could be had, trying in some way to produce partial, tentative answers, to the reworking of our environment and of ourselves by the fruit of this same scheme. That is the transition we are in; I don't think it is going to take very much longer. It will take some time. We are still in transition.

I should set out again to calibrate myself; I was very much moved recently by hearing Roger Revelle, just back from India, when he discussed the terrible prospect which he saw facing the Indian subcontinent because of the unprecedented failure of the southeast monsoon rains of the summer of 1965. This seems to have cut the Indian crops by as much as 15 per cent; he took a most pessimistic view of the circumstance. He felt it would inexorably mean the widespread coming of famine, genuine famine, in many parts of India. He saw it irremediable by any technical means at our present disposal. Even if the United States were willing and able to act promptly, it could at best alleviate and not prevent famine. That shows how narrow our technical base is as yet. Nevertheless, for our own community, for our United States, granted many difficulties of social organization, the fact of the matter is that we have resolved most of the problems of domestic primary production. For the world, we have not. The transition is really a transition, testing all nations to see whether they can expand this concern and this ability over all the face of the world. Maybe we cannot do it; but at least this is the century in which it is going to be tried. We note evidence of that from every newspaper headline for the past twenty years.

Farming production is not gone but is going, going in the sense that while it will be present as the essential basis of life, more and more of it is turned into mechanical hands. The job of the farmer will be that of the supervisor of machines, or better the supervisor of biological enterprises, and not the single farmer looking at the sky, concerned about his crops, year after year, without substantial change. His wisdom is not the wisdom of the elders who have been through their dry years and know how to manage with the seeds of bitter melon; his wisdom will be the rationalized wisdom of experience and control in the hands of the biochemist, the soil technologist, and the rest. This is a fantastic change; we are by no means prepared for it.

I am saying that the very basis, the very assumptions on which we rest our ordinary discourse, our values, our concerns, is going to shift with that change. Once again, I am trying not to be normative about this. Where my sympathies lie can be seen, I think, from the incautious prophecies I have made, but I don't wish to be normative. I point out only that when our ancestors perforce adopted the way of life which depends upon scratching the soil, living in one place, and getting the fruit of the land, they destroyed a magnificent, a beautiful, and a heroic way of life. It was the life of the hunter, whose skill in the chase, whose intimate relationship with prey both as taker and as dependent, is so strong to anyone who has studied hunting cultures, known hunters, seen the great ethnic pictures done in our time, or looked marveling at the magnificent cave paintings of Lascaux. Something was destroyed, never to be had again. I claim now the machine, for better or for worse, has become the way of life. We will see our metaphors, our images, our concerns, our very beings changed in response to these new experiences.

Now I come to the heart of my subject. We regard education as somehow indoctrination or exposure to the body of received culture, impressing the ability to take part in its transmission. This too will clearly require change. By now it is a truism that nobody can read all the good books that are published each year. They cannot be physically accumulated, let alone absorbed by anyone. Even in a single field it becomes impossible. Of course the category of good or bad would require a little argument; even so, giving certain generous benefit to my colleagues, I have the impression that some of all the books published should be worthwhile. But I can't even keep up with a tiny portion of it.

Just to mention one example, there is the *Physical Review,* which is the archival journal of U.S. physicists. They tell me there are good papers (not that I read them anymore). It comes out every two weeks, an issue about an inch and a quarter thick, in quarto size, with closely spaced double columns of very condensed mathematical material. That would be a library book every two weeks—such a book as when I was a graduate student, I felt content to master most of in six months of hard application. It is perfectly clear that this is a specialized journal. There are now two parts, marked A and B. The journal will be divided into six parts next year. For even those parts, you know how it will go. Remember, I am not talking about the Chinese, who have not yet begun to write.

Clearly the implication is that of specialization, and the strictness of specialization. I don't want to go into this anymore; Professor Feigl did, and he is quite right about it. How can we cope with this? A form of specialization which I would like to speak most strongly against is one that is not mentioned in academic contexts. In the academy there is a specialization which I shall call the horizontal plane, that is, where the anatomist does not know what the zoologist is doing. Bad enough, but what I am concerned about is rather that there does not arise a *vertical* stratification, one very close to us now, where the majority of people, even in a wealthy country like our own, simply are apart from the intellectual issues of our time. Of course this was always the case; at best one had a few flowerings of felt unity, perhaps around the cathedrals of medieval times. We will do the world a terrible injustice if we save the power, particularly in science, for a mysterious elite, the consequences of whose actions more and more change the very way of life, while the wellsprings of action are more and more mysterious to the majority of the population. I do not think I am harsh when I say I describe most of us at the present time, even in this room. I certainly describe most Americans.

What lies fifty years ahead in the face of the kind of change I was talking about? Let me read a philosopher much belabored in education but still a profound source. I do not read from his work in education, but from a seminal work on aesthetics, called *Art and Experience,* by John Dewey:

> We inherit much from the cultures of the past. The influence of Greek science and philosophy, of Roman law, of religion having a Jewish source, upon our present institutions, beliefs and ways of thinking and feeling is too familiar to need more than mention. Into the operation of these factors two forces have been injected that are distinctly late in origin and that constitute the "modern" in the present epoch. These two forces are natural science and its application in industry and commerce through machinery and the use of non-human modes of energy. . . . It is one manifestation of the incoherence of our civilization produced by new forces, so new that the attitudes belonging to them and the consequences issuing from them have not been incorporated and digested into integral elements of experience.
>
> The problem is so acute and so widely influential that any solution that can be proposed is an anticipation that can at best be realized only by the course of events. Scientific method as now practiced is too new to be naturalized in experience. It will be a long time before it so sinks into the subsoil of mind as to become an integral part of corporate belief and attitude. Till that happens, both method and conclusions will remain the possession of specialized experts, and will exercise their general influence only by way of external and more or less disintegrating impact upon beliefs, and by equally external practical application.

That is the text I am going to discuss in what follows. The key problem in this transition period, I submit, is so to fashion the kind of education throughout the world, geographically, horizontally, and vertically, that we enrich the subsoil of the mind, not producing the showy flowers of this or that scientific development, but the very subsoil from which all other activity of mind will spring. Until we do that we will be faced with a split society, an unstable one, internally and externally.

I see only one general way to describe the kind of education in science. It may be that what I say has implications far beyond science, but those implications I have for others to draw. Only one way occurs to me, and that is that the style, the form, the set of values, the objectives of education from the youngest child to the retired person no longer engaged in economic life—education will have to spread in time as well as in depth in the population—that style will approximate the style we now associate in the best laboratories with scientific research. That means the student will be conceived of not as confronting a body of material to be appreciated, learned, or subsumed; he will be conceived of as a complex living organism to which some assistance may be given, for whom some familiarity may be gained.

Education will emphasize directness of experience. Not the textbook, but the text will be its source; this, of course, is not strange to literary scholars either. It will imply involvement, a subjective concern, the magnificent word *commitment,* intellectual commitment to something, and *that* is the most important thing. We study what is interesting enough to absorb us, not because it is suitable for general education but because we want to learn it. I am tired of discussing what is suitable for education. I think we no longer have time in this world to specify those bits and fragments of an extremely complex body of knowledge which is right for everyone to have.

Education will have a kind of intensity because it has involvement. It will have a specific quality; it will have a concreteness, a need for specific details, but details seen with interest. It will have a much greater component of self-choice. It will not draw back from facing fundamental questions. It will try hard to cope with the nature of evidence: How do you know what you know? In this respect I am touching upon the philosopher's problems, the third analytical portion of philosophy, the epistemological side, which is indispensable in the daily work of the scientist; even if he does not do it very learnedly, he has to cope with it in a practical, straightforward way. It will not shrink before novelty; it will emphasize mutuality, as every scientist emphasizes in his own work each week the absolute dependence of any worker in intellectual activities, of anybody trying to get new ideas, upon many other persons, so that nothing is gained by oneself. At the same time it holds an absolute individuality, in which it is not enough to accept that which you need because it is based on authority; you must grapple with understanding it yourself. No one can spare you that. This curious opposition of individuality and dependence is most characteristic of research work. It is quite foreign to the traditional elementary school. What I claim is that this dichotomy cannot much longer be tolerated. All of those things, then, suggest the sort of change I foresee.

Let me go further. It is perfectly clear that the characteristic mode of education broadly viewed up till 1865, not to specify any more closely than that, will not be the mode characteristic beyond 2065. I cut out our present transitional years because both modes are present. The early education emphasized symbol and the use of symbol because the use of symbol would be necessary for conventional and learned knowledge, which could be imparted to the young man or the young woman who was growing on the farm, who knew firsthand the full moon, the snow, the cattle, and the wheat, who knew how to pry up the log with a small branch. Lincoln would not have walked very far to see a demonstration of simple mechanics because he saw that all the time; but he was willing to walk those eight miles to return or to gain the book, for a book with its conventional, arbitrary collection of symbols, bearing just a little wisdom, was something he could not otherwise have.

Therefore, the introduction of the young to the symbolic possibilities, reading, writing, and arithmetic, was the hub. Then, in better schools, Latin and Greek formed the community of the learned. Especially for the gentlemen who operated the British system for centuries, they were the key to general education. It was a little bit of the *quadrivium* and the *trivium* translated into the language of early nineteenth-century Harvard. That was enough to turn out the polished gentleman who became, by and large, the school teacher or the minister, or perhaps a lawyer of the Massachusetts community. This would describe other universities; I do not think the pattern was less complete elsewhere.

Beginning about 100 years ago, very largely around the frontier experience which was characteristic of the Middle West 120 years ago and of the West seventy years ago, a new element arose, a spreading of the industrial life into this cloister, which was a vestige from the once purely nonsecular University of Paris of 1,000 years ago. The symbol demands a richer experience of the substance; it is no use to read if you have not the experience. It is no use to watch the learned lectures, with brilliant demonstrations on television —or, perhaps it is of some use, but not much use—if you yourself have never had a single chance to see such phenomena in a different context. The roots of the research instruments which extend our senses so extraordinarily lie in the senses themselves. The senses, too, are instruments. The glib assumption that that which is perceived by eye and hand is given, and everything else is added to it by experience, by inference, and by reason is false. We know from the work of the psychologist, from every good nursery school teacher, from mothers and children, that the child must in part *learn* to use his hands and his eyes. He builds assumption; he builds inferences on them in a complex and subtle way as he learns to do this. It is not quite easy to show this, but I want to make my argument explicit.

In some sense the experience of the scientist is the experience of every individual of the human race, extended very far and suffused with an adult's kind of reason. The scientist borrows from the philosopher, from the symbol, much more than the child in his naive experience possibly could. But that there is a disparagement here, I deny. The senses are themselves not free from illusion. Inference, too, is not free from illusion. Both are a kind of instrument, both are to be used together, with a more or less explicit theory of how the world works, a theory built up, refabricated, and reworked as time goes on. The physicist has gone perhaps further looking for atoms, but he looks for them in the same way in which as a boy he first begins to take apart an old clock. The fact that he can't see the gears anymore makes it harder, but he manipulates those invisible parts in very much the same way, facing the same kind of difficulties.

Let me give an example. When I first looked through a microscope, I stood in line with other sixth- or seventh-graders, and waited rather expectantly. It was a big thing—it was a real microscope; the teacher allowed me to look through it for about fifteen seconds at something she had prepared. There were other people waiting; even fifteen seconds times forty would be a delay of ten minutes. She could not afford more than that. What I saw was a very good view of my own eyelashes. Having looked through my microscope, I was now supposed to believe in microscopic pictures. Of course I was a good student, and I learned what the teacher told me, but I really had no idea how on earth those people ever discovered anything but their eyelashes. I submit that is the common experience of mankind: what you see in a microscope is your eyelashes.

The microscopic view of the world is an essential feature of modern science; I don't want to argue that. If there were no modern microscopes, there would be no modern science. I don't speak of any special resort but of the whole idea of the fabric of the world on an invisible scale. That idea can be attained by looking carefully through a cheap microscope for a little while, but not by looking at your eyelashes for fifteen seconds, even in the best Zeiss. What it does take is not a learned investigation of the names of the cells, or a discussion of how the stain brings out the mitochondria so you can call them the Golgi apparatus. That isn't the idea; the idea is to have some chance to see that this is the instrument, with its own logic, its own traits, and to look at it this way and that way. It does not literally "make things larger."

If we give ten-year-old children a chance to do that, they will do it very well. My hope and judgment is that this can make a sizable difference, not to the degree to which they learn answers in the books about what microscopic things really are or how to draw them, but to the degree of conviction, to the *productivity* that these ideas will have in their future intellectual life; that is the key point. We don't have to look through a microscope to appreciate that the microscope shows us the world under high magnification, in a very strange light, but we best get that sense, I believe, when we look through a microscope ourselves; then it is something rather familiar, not just self-consistent.

We learn this by looking at the work of the great founders of modern science in the seventeenth century. I know one example from Robert Hooke, that strange genius of London in that same century. He had two new demonstrations each month for the meeting of the Royal Society, and they were magnificent. Almost anyone of them would have won a Nobel prize, so to speak, in the contemporary world; yet they thought this was nothing. He was a paid demonstrator, and it was up to him to do it. And he did it. Of course it was easier in those days, I will admit. His great book *Micrographia* was the first exploration in Europe, in popular language, of the microscopic world. He says all one needs is the microscope, "a faithful eye, and a sincere hand to record the things themselves as they appeared." And he tells how to do that: he says, don't look in just one light, move the light around; try one angle and then another. Pretty soon you will be able to reconstruct reality.

Note that it is not the given thing of the textbooks. (I would like to enter into the Feigl-Feyerabend argument concerning the roots of science because I think there a good deal was left out.) The textbooks leave the impression that the microscopic picture is something given. In fact it is not given. As long as the textbook is used, page 32 gives the illustration. I can learn that. But if I try to make that illustration, to produce it for myself from the material which is as hard for me to do as it was for the experts who made it, matters are different. Those experts struggled to get the lighting right, to get the material right in the first place; they had to find the right cell in the right condition, not two cells one on top of the other, not a bit of dirt that happens to come on the slide—dirt always does appear on the slides—not many other things. Theory always requires the right condition, but the issue is: What *are* the right conditions?

In short, it is all very easy to argue from given things, but science is characteristically not that which is given but that which is wrested somehow from a complex buzzing environment. It is not a logical action either; it is not logical for it demands manipulation of the world, changing it in many ways—I feel that unless this kind of understanding is at the back of my science, what I learn is at best a plausible mythology. Perhaps it is as good as Aristotle's, but I don't think it is any better. In fact it is rather less well argued because Aristotle foresees objections to every point. Nowadays these arguments go smoothly via the textbooks. Therefore I feel strongly that the kind of education we must look forward to in science will emphasize the nature of evidence. The nature of evidence is not mainly logical; the nature of evidence is chiefly experimental; that does not mean logic is separated from experiment. We cannot filter out that which is the reasoned and that which is the empirical. It is not that so much is given nor that equal masses are given. We had to find out if equal arms balance. The iron and sulphur which Professor Feigl was discussing is all very well, but I wonder if he has tried it? It turns out they usually don't make a combination of equal proportions. It works sometimes just as in the form in the book; yes, it works about seven-tenths of the time. We ought to face that. Dismiss it as experimental error? Yes, provided we know the answer.

I myself took part in a famous experiment which is familiar to every school child, in the British Commonwealth especially, and, I think, to not a few in our own circle. A candle is burned in water over an enclosed glass and the air changes, so that the water rises up. The volume change is measured. It comes out 15 to 20 per cent; this proves, it is said, that the candle will use up all the oxygen and go out. The true quantitative statements that most people with any kind of education can make of the structure of matter are that water is H_2O and, second, the air is 20 per cent oxygen. If we know nothing else about oxygen or per cents, we know that air is 20 per cent oxygen. It is very commonly taught, and it is even true—which gives it an advantage over many well-taught points. But it is not at all demonstrated by that experiment! We showed that in several different ways. We still got repeated objections on publishing this. How can we attack 100 years of scientific demonstrations in classrooms? But they were wrong, in fact. I am sorry to say it, but that is the fact of the matter. The water rise does not depend only upon burning up the oxygen. First, the candle goes out while the oxygen is still half there; it is the rate of burning that affects the candle. When the air is eight per cent oxygen, it usually stops and goes out. Nevertheless the water goes up 20 per cent. Why? Because the rise of the water level has to do with the heated air which expands to drive out water from the tube, and so on. Normally the candle does go out while there is still plenty of oxygen left. The group working on this for elementary education put white mice into the jar. These mice remain happy, and they try to eat the extinguished candle for the most part. In the classic theory, I suppose, this proves that mice can do without oxygen.

The lesson of this is very clear. People did that experiment, not because it was an honest experiment but because they were required to do it. Look at those complexities; it sounds so very, very simple. But the man who invented that experiment (I don't know who it was), about 100 years ago, did so on the basis of a very long series of gas experiments. When he set it up, he had thought through all those things that were needed to do the experiment right. But by elision of all these details and concerns, the sense was lost. Those who did the experiment never asked themselves: Is the oxygen all gone? What do you do to test that? You make a control; that the oxygen is *not* gone is the first thing you learn. Second, what are those variations that you find in any experiment? If the experiment is done on a layer of clay, for example, the water doesn't rise enough to speak of because the clay seals the tube off. The point is that it isn't simple to understand even such a little system as that, if we are honest about it, as an observation. Of course we can inflict it on the memory and then the examination asks, "Oxygen is . . . per cent of the air; 2, 20, 100?" Fill in box number 3, get the right answer, go to Harvard University. This is a caricature of our time, but it is not a broadly drawn one. The fact of the matter is that almost no experiment is so easy. Experiments are devilishly hard to perform.

When I was in school, I heard philosophers say (and they still tell me, I am sorry to say) that the thing to do is confront them with the experiment and then you know the experiment is always the ruler. I believe in that as a general principle. I suppose it would be treason not to believe that. I do believe it. But the fact is that in a seminar in contemporary physics, discussing contemporary problems—I am sure it is true in biology, astronomy, physics, and elsewhere

—there are always experiments that contradict theory, and the theory is still good. Now why is that? That is because we botched the experiment. Why did we botch the experiment? Because we botch nearly every experiment. Experiments are extremely hard to make. The experiment of the candle is the map of any kind of experiment. The world doesn't give itself as a logical connection; the world gives itself as the real world, with short circuits, air leaking in, candles that burn only half the amount of oxygen, and things of that sort. That is not all stipulated in the conditions. So we go and confront an experiment over and over until it is really well done, analyzed to the hilt. Now, this kind of involvement in experimentation can be secured at a simple level in candle-burning or in looking through microscopes, but it cannot all be done from books, never for any generation. Even what I say about the candle can be written down in a book and still fail to make its effect. I am not sure that anything can make the impact fully, but I am quite sure that only a direct involvement in the matter will work.

I am still an educational optimist. Almost all children before they come into the school system learn the most difficult, the most subtle, and the most useful of all intellectual skills: how to speak the mother tongue. If they had to be taught that in school, I hesitate to think what would happen to the world. Now, that language which they learn (I take it from the common phrase) from mother, and also from their peers, is clearly the most important of all learning. It is a symbolic study, but it is learned in a social context, taught with a great deal of kindness; it is a learning which is very favorable in teacher-student ratio, a learning which they gain free from examinations. Everyone who speaks a mother tongue knows that in the next thirty seconds he can quite easily utter a sentence which he has never uttered before, and he can understand sentences such as the one I am now saying which he has never heard precisely in the same way. We don't shirk at all before this novelty of recombination. If we don't know the language, we can't perform. I don't know German, yet I can say *Guten Abend;* I can do the same thing in French, and a little bit better in Italian; but I am unproductive in all these languages. I can barely get out my phrases, and then I am lost—I can speak only painstakingly and painfully. I think many of us have the same experience.

We learn our mother tongue in a different way, and we have a different kind of knowledge. What I claim as the desirable kind of learning is the productive kind of those who speak English sentences, who ask and understand new English sentences at any level. They don't need to be very fancy sentences; it is quite true that we can always learn more and more complex ways of talking about things. But the child looks out of the Berlitz ads and says he can speak better Ewe than others can—he must have learned this at an awfully good school because the number of Ewe speakers in America is very small. Yet a five-year-old can do it! What is the reason for this?

I concede that this argument is not without a hole, for there are important developmental changes which might affect it. Maybe children are much better at learning languages than they are ever again at anything else—there is some evidence of that; still, I don't think we can argue that all the other criteria are unimportant. I maintain there are many reasons for it. Intense involvement, essential reward in terms of time, a detailed study, no examinations. The emotional tone in which language is learned is extraordinary. A child learns those dirty words when he gets mad at his peers in the street and they beat him; he learns lullabies and prayers; he learns formal statements which he must repeat precisely on command; he is freely given playful trivia. All this is for five-year-olds. It is perfectly true that all these things are the case; I doubt very much if any single person who has ever learned a language as the mother tongue has not had that kind of experience. People who learn languages well, who gain a productive knowledge, have put themselves into a context of speaking it and using it in many different ways.

This is quite unlike the way we follow in science. Science is pursued, mostly since the nineteenth century, with a great air of austere and forbidding perfection. We speak of *the exact sciences*. We see the delicate brass instruments under glass jars, the austere laboratory, the scientist either in a black tail-coat (that is the way he was a hundred years ago) or a white smock (that is the way he is now). Both of these things are outrageous theater, a remark which is verifiable by looking at any laboratory. The emotional tone in which the school makes science, not so much the way the scientists make science, is the restrained emotional tone that I have just described. I have nothing against that, for I am an admirer of the austere, the sparse, the spare, the functional; but I submit that to do everything in one emotional tone for as complex and as deep a goal as the advance of education in science is a grave mistake, a grave error. We can be pretty sure of that. I suspect that the extraordinary discrepancy between the number of women and the number of men who enter science at any level has a very great deal to do with this demanding emotional tone in which science is set. That tone has absolutely nothing to do with most of science. It is generally belied by the conduct of the researcher. The jokes and frivolity that can surround a synchrotron do not enter the classroom. But this is absurd. Surely if there is any need for this kind of relief, it will be present for the youngest children. The more professional we get at making a living, the more we seek a long-range goal and have an overwhelming motivation—the less we would need the kind of humanism, in a narrow sense of the word, which surrounds most successful laboratory enterprises.

Again there is a wide spectrum, and again I would not insist on a fixed pattern. There are very successful laboratories and scientists who still work in this austere frock-coated fashion; it is their privilege. There are some people who act that way all through life. I have nothing to say for or against it. I am not judging among persons; but I am saying that society contains persons of many kinds, and to impose the feeling that there is a special personal tone that must go with a certain intellectual activity is to bar that intellectual activity from the greatest portion of the community.

I have an impression which I could not document: that the nineteenth century is responsible for this; it rises together with utilitarianism in the early part of that century. Mr. Bentham, whose skeleton hangs so austerely in the lecture room in London University, made it for us. It does not go at all with the seventeenth-century masters of the beginnings of modern science. Vesalius, in the great plates of his *Fabric of the Human Body,* was not above putting behind the successive figures of the cadaver the Tuscan landscape going through all four seasons. We are in summer with the full man; by the time we get to the bony skeletal remains, there is behind them only the bleak landscape of winter, with all the trees bare, the skeleton of the trees showing behind the skeleton of the man. Now I don't want to overemphasize this case; it is not my view that just adding artistic strength to illustrations in scientific books will succeed, nor is it my view that we all can be as fine artists as those engravers were, nor is it my view that the scientific world must learn to recoup the naive and fresh quality of its inventors.

It would be nice to have some of those things back, but my point is more strongly based, that is, unless we extend ourselves to *include* this tone somehow, in all the ways and variance it can have in the modern world, we are lost. We will not succeed in making science part of the subsoil of the mind; we will divide because the nonscientist will be repelled by the technical. The division will grow and grow. There are terrible dangers which a split society faces, which no institution of learning, especially none devoted to the moral unity of man as is this one, could tolerate. We must try to find ways. Not to take the dry, the austere, out of science—that is not my idea—but to add the enriched to it. That is closer to what I want to say.

Finally, I must in this context touch on another point. In the end, the practitioners of science, on the level of research, work at it not only because it is a good job, not only to excel, not only because it is a logical puzzle, but also for inner reasons, for aesthetic ones. A very similar instinctive aesthetic reaction is shown by most people to many things which partake of the laboratory. The physicists I know quite well who have those other motives that I mentioned respond in a similar way to things like the Super-ball, to puzzles which appear in the magazines, and to a thousand other toys and amusements. I do not think these matters are separate. The playful element is indispensable to most intellectual activity, a recombining without the constant demand that each pattern can be functionally meaningful. Can the painter or the poet, or indeed the scholar, honestly say to himself he has not done something of that kind? In his reformulations, in his prophecies, though not necessarily in the final form after a careful filtration process has occurred but in the formulation, there is held the very recombination we call play. What is play as we see it in animals? There is no question open about the play of the cat, though it is a profound subject; and I admit that I don't know much about it, but I myself can see in the play of the cat the preparation for the hunt. It is not meaningful in itself, but it is a true preparation for the hunt. I see in the play of human beings the hunt for novel ideas, the hunt for order, the hunt for tests of their own strength. If we restrict these, if we take these out of education, or if we take these out of schools, we commit a grave error. We must use play.

I should like to urge as well that the context in which we study and teach science—and by implication, though I can't speak with authority, many other topics—be widened considerably. The world is full of persons who find themselves threatened by boredom. It is a characteristic of the modern world which has turned away from primary production, though I must recall that the Indian village, while very busy in the right season, is in the off season very idle. Men always have a problem of this kind; but our modern problems are special. Take air flight, which is a small experience but an experience in which many influential persons engage in our society. I notice my fellow passengers very frequently spending their time in pursuits not commensurate with their abilities, neither playfully nor interestingly. I would like to see as a mere example—one that could be multiplied a hundred times throughout the whole society—an effort made to produce in him who runs some reading of the world, reading not just books, not just James Bond, but reading the variety of meteorological, geographical, sociological and physical phenomena in the country which unrolls below the airplane. The world below in which we live, whether it be clouds or whether it be crop land, is legible. If I sit next to an expert who is articulate in any of these subjects, he can spin a fascinating story out of it. Physicists I know may be navigating the airplane, or they hold a hanging watch or a tag on a piece of string and watch it swing back and forth as the airplane accelerates and decelerates. It is in microcosm the launching of a rocket.

What I am saying is that in our world we have split ourselves off somehow. There is no rich context of intellectual involvement on a simple, playful level; that is why we do not yet have a subsoil fertile for scientific ideas. If we had, then everyone would possess great powers in one way or the other. Of course not all need be the same; it is not that everyone must study Accelerometry I and II so he can ride in airplanes; that is not right. But why don't we have a thousand booklets inside the airplanes, little ideas, cutouts, gadgets for play of all sorts. Why isn't this extended to playgrounds, to every other leisure activity? We do not have a spate of these things demanded by people who know this is an interesting and a valuable thing to do, who enjoy it, who simply take pleasure in the small notions of science as well as the largest ideas of cosmology. Both of these extremes should be present; in short, until this is improved we will not see a world in which science is in fact at home in intellectual life. Until then its symbols, its metaphors, its full community will be absent from everyday living.

I submit this as a key question of the next 100 years in education: to try to produce such watered subsoil that will take its form in every diverse fashion. It will also include (perhaps it is obvious in what I have been saying) a speaking out for a kind of amateurism, a widespread amateurism in science as there is a widespread amateurism in literature. For what "serious readers" do the novelists write? It is true that many of them write for other novelists, and for a few critics, but it is generally held that that is not the wisest way to maintain a living literature. The world provides raw material for the novelists, but it provides as well the raw material of science. The world is there with us; we cannot get away from it, even locked in the isolation cell, we still have it. This is the kind of attitude to which I hope schools will direct themselves.

Every productive speaker of the English language, that is, every man on the street as well as every research scientist, knows that he is learning and changing and formulating; he is constantly studying; he is learning new things from the newspapers. Study is a small matter in language, but that is why language is productive; that is why people can read and repeat, combine and form new words, and have new ideas. Now it is not easy to carry through these strictures. I have not talked about it in enough detail to let you see how it might be done in the context of schools or out of schools: how it might be done for very young children or how it might be done for people long removed from school years. I think the time scale of our consideration is so long that I need not talk about our contemporary problems and the efforts to try to solve them. Surely they are very numerous, and surely there are many better ideas than those we have seen yet; but I still think the direction and the burden of what I am saying will by repeated examples become somewhat clear.

I come now to that unity which has been the implied concern of many speakers against the cleavage of our culture. It is evident to all of us, as I began by saying, that unity will never be complete. Specialization is indispensable as knowledge is overwhelming. But there is a kind of unity, a unity of method and involvement, which is wider than a mere unity of the subject of discourse. There are many activities of men and women—the activities of the devoted teacher of Latin in the secondary schools, those of the kindergarten teacher, the ballerina, the good fisherman, the lithographer, the scientist—which are all characterized by a pursuit, a kind of fire, a kind of spark, a kind of concern for the consequences themselves, not merely for the fulfillment of an indispensable requirement of the community. The parliamentarians of the nineteenth century in Britain learned Latin and Greek; as far as I know, no single good Latin poem was ever produced by any of those statesmen who for two centuries studied how to write Latin poetry in the best schools of England. I don't doubt there was something gained by that, a kind of community. They knew that if they made their famous "false particle" in the House of Commons, the House would laugh. Well, that was desirable. But although it may still be desirable today, it will less and less be true in the future that that sort of community can be attained by any important fraction of the population. That is simply because the histories of the world and the range of traditions coming into contact today are simply much too wide.

I observed from the very beginning that I am talking

about 100 years from now when the contribution to any particular sphere of intellectual life will be as great in Chinese, or greater perhaps, than in any other tongue, and of course it will not be lacking in the English language, Spanish, or in Ewe for that matter. This kind of world is more than our scholars could cope with. It was not the world seen by the scholars of the past; it could not in the nature of things be so. An educated man knew Latin, and 500 years ago it was enough if he spoke a vernacular or two. Then French: of course everyone knew French, and wherever there was a very well-educated man from the small countries of Europe, Scandinavia, the Lowlands, or even old Russia, such a man spoke beautifully four or five languages. I do him all justice, but even that can't begin to span the languages of man. Esperanto was a "universal tongue," built on the Romance languages. Obviously, a Japanese or Chinese would have a very hard time learning that! This is not the style of the world ahead; our world is by far more open. Our world is going to demand, I think, a community founded on other things.

It seems to me it can be founded, as much as on anything, on the sense of novelty, of deep involvement, on those emotional and methodological tendencies that we gain from experience with the material itself. This is one role which science can play because science, more than the rest of culture, transcends tradition; science is found in all countries, in all ages, in all walks of life. The world will respond to us, give some reward, or reinforce a correct prediction whenever we bounce a ball or point a tube—whether we come from the right social class, speak coarsely, use an improper accent, bear a scar on our face, or whether we are generally a bright and promising child or a bad and disappointing one. There is in the physical world a guaranteed reward for the virtue of truth; opportunity we owe to all men. An educational system in this world, so wealthy, so threatened, so plainly in transition, will have to build upon that. Somewhere in this region lies the widest community of the future. I could summarize it in one sentence: that community will one day lie—I do not hesitate to use the right word—in a common delight in the progress of the intelligence of men.

CONTRIBUTORS

Ludwig F. Audrieth, born in Vienna, Austria, in 1901, is late professor emeritus of chemistry at the University of Illinois. He was science attaché at the American embassy in Bonn and a consultant on science to the State Department for some years. Coauthor of *The Chemistry of Hydrazine* and *Non-Aqueous Solvents,* his main interest in recent years is the topic of his paper here.

Milton Burton, born in New York in 1902, was chief of the radiation chemistry section of the atomic energy project at the University of Chicago, 1942–1945, and is now professor of chemistry at the University of Notre Dame as well as director of its Radiation Laboratory. Coauthor of *Photochemistry and the Mechanism of Chemical Reactions,* he has been a Fulbright lecturer and Guggenheim Fellow.

Michael J. Crowe, born in Minneapolis, Minnesota, in 1936, is assistant professor in the General Program of Liberal Studies at the University of Notre Dame. He holds a Ph.D. degree in the history of science from the University of Wisconsin and specializes in the history of nineteenth-century physical science and mathematics. He is the author of *A History of Vector Analysis.*

Farrington Daniels, born in Minneapolis, Minnesota, in 1889, is now emeritus professor of chemistry at the University of Wisconsin and research associate of the Solar Energy Laboratory. Author among other books of *Mathematical Preparation for Physical Chemistry* and *Chemical Kinetics,* he has been vice-president of the National Academy of Sciences and president of Sigma Xi and the American Chemical Society.

Herbert Feigl, born in Austria-Hungary in 1902, is professor of philosophy at the University of Minnesota and director of the Minnesota Center for the Philosophy of Science. A prolific author, he is co-editor of the influential series, *Minnesota Studies in the Philosophy of Science.*

Erwin N. Hiebert, born in 1919 in Saskatchewan, Canada, is professor of the history of science at the University of Wisconsin. He is the author of *The Impact of Atomic Energy* and *Historical Roots of the Principle of Conservation of Energy.* His interest has been the history of physical science and thought since 1800, with special emphasis on the history of thermodynamics.

Richard McKeon, born in New Jersey in 1900, is Distinguished Service Professor of Greek and Philosophy at the University of Chicago. Author of *Freedom and History* and *Thought, Action and Passion,* he has been United States delegate to UNESCO, president of the International Institute of Philosophy and of the American Philosophical Association.

Philip Morrison, born in New Jersey in 1915, is now professor of physics at The Massachusetts Institute of Technology. He was physicist and group leader at Los Alamos, and his recent research interests have been in the borderline areas of astronomy and particle physics. He is actively associated with Educational Services Incorporated, whose programs for the revision of science courses in secondary schools have been widely adopted.

Elizabeth Sewell, professor of English at Fordham University, has published a collection of poems as well as a number of novels and literary studies. An Englishwoman, she is the holder of a Ph.D. degree from Cambridge University in England and an honorary D.Litt. from St. Peter's College, Jersey City. Among her books are *The Human Metaphor* and *The Orphic Voice.*

John E. Smith, born in Brooklyn in 1921, is professor of philosophy at Yale University. Author of *Royce's Social Infinite, Value Convictions and High Education, Reason and God,* and *The Spirit of American Philosophy,* he has been Dudleian lecturer at Harvard University and visiting professor at the University of Heidelberg.

INDEX

Agassiz, Louis, 10, 116
D'Alembert, Jean Le Rond, 29
Alice's Adventures in Wonderland, 3
Andrews, Thomas, 113
Ångström, Anders Jonas, 109
Appleton, William Henry, 73
Aquinas, Saint Thomas, 83, 86
Aristotle, 41, 83, 133, 150, 229
Armstrong, William George, 66
Arrhenius, Svante, 77, 97
Art and Experience, 223
L'Assmmoir, 8
Auguste Comte and Positivism, 27
Augustine, Saint, 170, 172
Avenarius, Richard, 49
Avogadro, Count Amedeo, 110

Babbage, Charles, 28, 108, 122
Baeyer, Adolph von, 178
Bain, Alexander, 48
Bancroft, Wilder, xix
Barker, George Frederick, 76
Barzun, Jacques, 122
Beltrami, Eugenio, 107
Bergson, Henri, 11, 97
Bernard, Claude, 120, 121
Bernard of Chartres, 105
Berthelot, Pierre Eugène Marcelin, 110, 113
La Bête Humaine, 8
Blake, William, 13
Bohm, David, 149

Bohr, Niels, 110, 140
Boltzmann, Ludwig, 63, 139
Bolyai, Johann, 106
Boole, George, 48, 108, 122
Born, Max, 148, 149
Bosch, Karl, 179
Bourget, Charles Joseph Paul, 6
Boussingault, Jean Baptiste, 121
Braithwaite, Richard Bevan, 137
Brave New World, 7
Brentano, Franz, 144
Broglie, Louis Victor de, 149
Brown, Robert, 118
Brown-Séquard, Charles Édouard, 121
Brücke, Ernst Wilhelm von, 119
Buffon, Georges Louis Leclerc, 175
Bulletin of the Atomic Scientists, 204
Bunsen, Robert Wilhelm, 109
The Burning Glass, 6
Bush, Vannevar, 193
Butler, Samuel, 11
Butlerov, Aleksandr Mikhailovich, 110, 111

Campbell, Norman Robert, 137
Cannizzaro, Stanislaus, 110
Cannon, Walter F., 122
Cantor, George, 106, 108
Čapek, Karel, 7
Carnap, Rudolph, 137, 138, 147
Carnot, Sadi, 63, 93

Carpenter, William Benjamin, 67, 68, 69
Carus, Paul, 77
Cauchy, Augustin Louis, 114
Cellular Pathology, 118
Chaplin, Charlie, 7
Chemical Abstracts, xx, 193
Chesterton, Gilbert Keith, 98
Clairault, Alexis Claude, 29
Clapeyron, Benoît Paul Emile, 63
The Classification of the Sciences, 48
Clausius, Rudolph, 63, 64, 93, 111, 112
Clifford, William Kingdon, 107
Cohn, Ferdinand Julius, 120
Colding, 63, 97, 112
Coleridge, Samuel Taylor, 13
Collingwood, R. G., 16
Compton, Arthur, 204
Comte, Auguste, 28, 34, 35, 39, 40, 42, 43, 44, 47, 48, 49
The Correlation and Conservation of Forces, 73
Correlation of Physical Forces, 66
Cours de Philosophie Positive, 28, 43
Cusano, Nicholas, 219
Cuvier, Georges, 10, 11, 13

Dalton, John, 110, 123
Daniels, Farrington, xii
Dante Alighieri, 13
Darwin, Charles, 11, 73, 115, 116, 117, 156, 160, 170
Davaine, Casimir Joseph, 120
Davy, Sir Humphry, xix
Dawes, Ben, 122
Decline of the State of Science in England, 28
Dedekind, Julius Wilhelm Richard, 108
De la Mare, Walter, 6
Democritus, 110, 133

De Morgan, Augustus, 48, 108, 122
The Deputy, 9
Descartes, René, 29, 32, 33, 34, 169
Descent of Man, 116
Deville, Henri Étienne Sainte-Claire. *see* Sainte-Claire Deville, Henri Étienne
Dewey, John, 223
Dilthey, Wilhelm, 49
Dioptrics, 32
Le Disciple, 6
Discours de la Methode, 32
Discourse on the Study of Natural Philosophy, 28
The Doctrine of Uniformity in Geology Briefly Refuted, 117
Doppler, Christian Johann, 109
Du Bois-Reymond, Emil, 141
Duhem, Pierre, 91, 92, 93, 94, 95, 100

Eddington, Arthur, 87, 88, 142, 146
Einstein, Albert, 63, 115, 122, 134, 142, 143, 144, 149, 165, 172, 173, 174
Electricity and Magnetism, 31
Ellison, Ralph, 7
Encyclopedia Britannica, 112, 115
Erasmus, Desiderius, 11
Euclid, 124, 137
Euler, Leonhard, 29, 113
Eureka, 5
The Evolution of Physics, 115, 122
An Examination of Sir William Hamilton's Philosophy and of the Principal Philosophical Questions Discussed in His Writings, 27
Experiments and Observations on Electricity, 176
Exposition du Systéme du Monde, 30

Index

Fabric of the Human Body, 235
Faraday, Michael, xix, 31, 114, 122
Fechner, Gustav, 78
Feigl, Herbert, 217, 222, 229, 230
Feyerabend, Paul K., 138, 139, 140, 229
First Principles, 69
Fizeau, Armand Hippolyte Louis, 115
The Flashing Stream, 6
"Form and Content," 146
Foster, Michael, 168
Franklin, Benjamin, 176
Fresnel, Augustin Jean, 113, 114
Freud, Sigmund, 9, 97, 170

Galilei, Galileo, 33
Galois, Evariste, 108
Galton, Sir Francis, 118
Gassendi, Pierre, 13
Gauss, Carl Friedrich, 106, 107, 113
Gegenbaur, Karl, 119
"The Genesis of Science," 48
Geological Evidences for the Antiquity of Man, 116
Geometry, 32
Germinal, 8
Das Gesetz von der Erhaltung der Kraft und Seine Beziehung zur Metaphysik, 85
Gibbs, Josiah Willard, 139
Gilbert, Sir William Schwenck, 8
Das Glasperlenspiel, 11
God and the Astronomers, 89
Gödel, Kurt, 141
Goethe, Johann Wolfgang von, 5
Golgi, Camillo, 228
Grassman, Hermann Günther, 107
Gray, Asa, 116
Greeley, Horace, 73
Green, George, 113

Grove, William Robert, 66, 69, 70
Guldberg, Cato Maximilian, 113
Gutberlet, Konstantin, 84
Gutenberg, Johann, 216

Haber, Fritz, 179
Haeckel, Ernst, 72, 73, 85, 116, 141
Haldane, John Burdon Sanderson, 86, 87, 88
Hamilton, Sir William, 27, 28, 34, 35, 39, 40, 43, 47, 48, 49
Hamilton, William Gerard, 107
Hankel, Hermann, 107
Heat as Mode of Motion, 112
Heidegger, Martin, 51
Helmholtz, Hermann, 63, 70, 72, 73, 82, 107, 111, 112, 120, 121
Hempel, Carl Gustav, 137
Hereditary Genius, 118
Hermite, Charles, 108
Herschel, Sir John Frederick William, 28, 29
Hersey, John, 9
Hesse, Hermann, 11
Himmelfarb, Gertrude, 122
Hiroshima, 9
Hiroshima mon Amour, 9
History of Scientific Ideas, 37
History of the Inductive Sciences, 28, 37
Hobart, R. E. *see* Miller, Dickinson S.
Hochhuth, Rolf, 9
Hoek, 115
Hooke, Robert, 118, 228
Hooker, Sir Joseph Dalton, 116
Horstmann, 113
Hoyle, Fred, 7
Huggins, William, 109
Hugo, Victor, 3, 5, 12, 13, 14
Hume, David, 149, 150
Husserl, Edmund, 49
Hutchinson, Evelyn, 15
Huxley, Thomas, 70, 73, 116

Ihde, Aaron John, 122
In Good King Charles's Golden Days, 11
In Memoriam, 3
Infeld, Leopold, 115, 122
Inge, William Ralph, 88, 89, 90
International Scientific Series, 73
Introduction to the Study of Experimental Medicine, 121
Invisible Man, 7

James, Henry, 98
James, William, 78, 98
Jeans, Sir James, 87
Jefferson, Thomas, 175, 176
Jenkin, Fleeming, 117
Jevons, William Stanley, 108
Jordan, Camille, 108
Joule, James Prescott, 63, 70, 97, 112
Journal of the American Chemical Society, xix, xx
Jung, Carl Gustav, 97
Jussieu, 10

Kafka, Franz, 9
Kant, Immanuel, 27, 40, 49
Kekulé von Stradonitz, Friedrich August, 110, 111
Kelvin, William Thomson, 63, 97, 111, 112, 114, 117
Kirchhoff, Gustav Robert, 109
Klein, Felix, 108
Koch, Robert, 120
Kölliker, Rudolf Albert von, 117, 119
Kowalewsky, Alexander, 116
Kronecker, Leopold, 108
Kulturwissenschaft und Naturwissenschaft, 49

Lagrange, Joseph Louis, 13, 29, 30, 31, 113
Lamarck, Jean Baptiste, 11
Lamartine, Alphonse, 10
Laplace, Pierre Simone de, 29, 30, 31, 113
Lavoisier, Antoine Laurant, 122, 175
La Légende des Siècles, 12
Leibniz, Gottfried Wilhelm von, 13, 32, 33, 122, 169
Leo XIII, Pope, 82
Leucippus, 133
Lewis, Clive Staples, 7
Lie, Sophus, 108
Liebig, Justus von, 68, 73, 179
The Limits of Concept Formation in the Natural Sciences (Grenzen der Naturwissenschaftliche Begriffsbildung), 49
Lincoln, Abraham, 225
Liouville, Joseph, 108
Lister, Joseph, 120
Livy, Titus, 105
Lobachevski, Nicholaus, 106
Locksley Hall Sixty Years After, 6
Lodge, Sir Oliver, 85
Logic: Deductive and Inductive, 48
A Logical Introduction to Historical Science, 49
Longfellow, Henry Wadsworth, 10
Loschmidt, Joseph, 112
Ludwig, Carl Friedrich Wilhelm, 121
Lunn, Arnold, 86, 87, 88, 89
Luther, Martin, 89
Lyell, Sir Charles, 116

MacCullagh, 114
Mach, Ernst, 30
Magendie, François, 120, 121
Maher, Michael, 83
Mallarmé, Stéphane, 6
Malthus, Thomas Robert, 182
Man in Thought, 14
Man's Place in Nature, 116
Margenau, Henry, 137

Index

Marshall, Charles Burton, 183
Marx, Karl, 9
Maxwell, Grover, 146
Maxwell, James Clerk, 24, 31, 88, 107, 111, 112, 114, 115, 122
Mayer, Julius Robert, 63, 70, 72, 97, 112
Mécanique Analytique, 30
Mécanique Celeste, 30
Mendel, Gregor, 118
Mendeleev, Dmitri Ivanovich, 110, 111
Méray, 108
Mercier, Cardinal Désire, 82, 83
Merz, John Theodore, 111
Meteors, 32
Meyer, Lother, 110, 111
Meyerson, Emile, 97, 99
Michelson, Albert Abraham, 115
Micrographia, 228
Mill, John Stuart, 27, 28, 33, 34, 35, 36, 37, 39, 44, 46, 47
Miller, Dickinson S., 149
Mind, 168
Minkowski, Hermann, 143
Modern Times, 7
Der Monismus als Band zwischen Religion und Wissenschaft, 72
The Monist, 77
Morgan, Charles, 6
Morley, Edward Williams, 115
Morley, John, 59
Müller, Johannes, 120
Murray, John, 115

Nagel, Ernest, 140
"Naming of Parts," 8
Naudin, 118
Needham, Joseph, 14, 15
Nehru, Jawaharlal, 185
New System of Chemical Philosophy, 110
Newton, Sir Isaac, 13, 28, 29, 30, 33, 34, 36, 105, 112, 115, 122
Nietzsche, Friedrich Wilhelm, 97, 133
Nieuwland, Rev. Julius Arthur, 181
1984, 7
Northrop, Filmer Stuart Cuckow, 137, 146, 147
Novum Organon Renovatum, 37, 38

"On First Reading Some Modern Physicists," 18
On Learned Ignorance, 219
On the Philosophy of Discovery, 28, 38
Open Court, 77
Organic Chemistry Based on Synthesis, 110
The Origin of Species, 8, 115, 116, 117
The Orphic Voice, 19
Ostwald, Wilhelm, 77, 78, 79, 81, 82, 96, 97, 100, 179
Owen, Robert, 117

Paradise Lost, 9
Passage to India, 3, 11, 15
Pasteur, Louis, 120
Peirce, Benjamin, 107
Peirce, Charles Sanders, 108
Penal Colony, 9
Pepper, Stephen Coburn, 146
The Philosophy of the Inductive Sciences, 37
Physical Review, 222
Plato, 168
Poe, Edgar Allan, 5, 7
"Poem Feigned to Have Been Written by an Electronic Brain," 8
Poincaré, Henri, 13, 142, 146
Polanyi, Michael, 4, 14
Popper, Karl Raimund, 138, 139, 147

Index

Popular Science Monthly, 73
Princess, Ida, 8
Principia, 30, 112
Principia Mathematica, 18
The Principles of Empirical or Inductive Logic, 49
Principles of Sociology, 71

Redi, Francesco, 120
Reed, Henry, 8
Reichenbach, Hans, 135, 137
Resphigi, 115
Revelle, Roger, 177, 220
Rey, Abel, 91
Richards, I. A., 6
Rickert, Heinrich, 49, 50
The Riddles of the Universe, 141
Riemann, Bernard, 106, 108
Ross, Sir Ronald, 5
R.U.R. (Rossum's Universal Robots), 7
Royal Society, London, Proceedings, 176
Russell, Bertrand, 18, 97, 146

Sachs, Julius von, 121
Saint-Gilles, 113
St. Paul, 16
Sainte-Beuve, Charles Augustin, 10
Sainte-Claire Deville, Henri Étienne, 113
Schlegel, 107
Schleiden, Matthias Jakob, 118
Schlick, Moritz, 146
Schliemann, Heinrich, 4
Schopenhauer, Arthur, 144
Schrödinger, Erwin, 149
Schultze, Max, 119
Schwann, Theodor, 118
Science and Civilisation in China, 14
Science and History: A Critique of Positivist Epistemology, 50
"Science and Poetry," 6

Science and Religion, 173
La Science et l'Hypothèse, 13
Science et Méthode, 13
The Search, 6
Secchi, Pietro Angelo, 109
Sechenov, Ivan Mikhailovich, 121
Shakespeare, William, 3, 12
Shaw, George Bernard, 11
Shryock, Edwin Harold, 122
Shuster, George, xi
Singer, Charles, 122
Snow, Sir Charles, 6
Spallanzani, Lazzaro, 120
Spemann, Hans, 4
Spencer, Herbert, 48, 69, 70, 71, 73, 82, 96
Spengler, Oswald, 89, 100
Spinoza, Benedict, 169
Steinmetz, Charles Proteus, 196
Stevenson, Adlai, 190
Stokes, George Gabriel, 114
Summa Theologica, 86
Swift, Dean, xvi
Symbolic Logic, 49
Synthetic Philosophy, 71
System of Logic, 27, 28, 34

Tait, Peter Guthrie, 107, 112
Tennyson, Alfred, 3, 5
That Hideous Strength, 7
Théorie des Fonctions Analytiques, 30
Thompson, D'Arcy Wentworth, 15
Thomson, William. see Kelvin, William Thomson
Times (London, 4)
Traube, Ludwig, 113
Treatise on Natural Philosophy, 112
Tyndall, John, 97, 112, 120

Uzziah, King, 169

Valéry, Paul Ambroise, 6, 14

Index

La Valeur de la Science, 13
The Variation of Animals and Plants under Domestication, 117
Die Welträthsel (The Riddle of the Universe), 72
Venn, John, 49
Verne, Jules, 7
Vesalius, Andreas, 235
Vico, 33
Vigier, Jean Pierre, 149
Villemin, Jean Antoine, 120
Virchow, Rudolf, 118, 119
Virtanen, Arturri, 182

Waage, Peter, 113
Wain, John, 8
Wallace, Alfred Russell, 5, 116
Weber, Max, 50
Weierstrass, Karl Theodor, 108
Wells, Herbert George, 7
Weyl, Hermann, 142

Whewell, William, 27, 28, 34, 35, 37, 39, 47
Whitehead, Alfred North, 18, 97, 170, 174
Whitman, Walt, 3, 5, 11, 12, 14, 73
Wiesner, Jerome Bert, 176
Wilberforce, William, 117
Wilson, J. Walter, 119, 122
Wittgenstein, Ludwig Josef Johann, 100, 147
Wordsworth, William, 13, 19
Woronin, 121
Wroblewski, August, 77
Wunderlich, Carl August, 112

Youmans, Edward Livingston, 73, 74
Young, Thomas, 113

Zirkle, 122
Zola, Émile, 8, 9